CONTRACT ADMINISTRATION

FOR THE BUILDING TEAM

SEVENTH EDITION

D0279187

Other titles by the Aqua Group

*Pre-Contract Practice for Architects and
Quantity Surveyors*

Tenders and Contracts for Building

Fire and Building

CONTRACT ADMINISTRATION

FOR THE BUILDING TEAM

SEVENTH EDITION

THE AQUA GROUP

With sketches by
Brian Bagnall

OXFORD
BSP PROFESSIONAL BOOKS
LONDON EDINBURGH BOSTON
MELBOURNE PARIS BERLIN VIENNA

Copyright © The Aqua Group 1965,
1972, 1975, 1979, 1981, 1986, 1990

BSP Professional Books
A division of Blackwell Scientific
    Publications Ltd
Editorial Offices:
Osney Mead, Oxford OX2 0EL
25 John Street, London WC1N 2BL
23 Ainslie Place, Edinburgh EH3 6AJ
238 Main Street, Cambridge
    MA 02142, USA
54 University Street, Carlton
    Victoria 3053, Australia

First edition published by
    Crosby Lockwood & Son Ltd 1965
Second edition 1972
Third edition published by Granada
    Publishing Ltd in Crosby Lockwood
    Staples 1975
Fourth edition published by Granada
    Publishing Ltd 1979
Fifth edition 1981
Reprinted 1983
Sixth edition published by Collins
    Professional and Technical Books 1986
Seventh edition published by BSP
    Professional Books 1990
Reprinted 1992, 1994

Set by Kudos Graphics, Horsham,
    West Sussex
Printed and bound in Great Britain by
    Hartnolls Ltd, Bodmin, Cornwall

DISTRIBUTORS

Marston Book Services Ltd
PO Box 87
Oxford OX2 0DT
(*Orders:* Tel: 0865 791155
        Fax: 0865 791927
        Telex: 837515)

USA
Blackwell Scientific Publications, Inc.
238 Main Street,
Cambridge, MA 02142
(*Orders:* Tel: 800 759-6102
        617 876-7000)

Canada
Oxford University Press
70 Wynford Drive
Don Mills
Ontario M3C 1J9
(*Orders:* Tel: 416 441-2941)

Australia
Blackwell Scientific Publications Pty Ltd
54 University Street
Carlton, Victoria 3053
(*Orders:* Tel: 03 347-5552)

British Library
Cataloguing in Publication Data

Contract administration. – 7th ed.
    1. Great Britain. Buildings. Construction.
    Contracts. Administration
    I. Aqua Group
    692.80941

ISBN 0-632-02632-4

Contents

Introduction   vii

1 The building team   1

2 Placing the contract   12

3 Progress and site meetings   24

4 Site duties   33

5 Instructions and variations   47

6 Interim certificates   65

7 Completion, defects and the final account   82

8 Delays and disputes   96

9 Insolvency   110

Index   131

Introduction

Introduction

It is over a quarter of a century since the formation of the Aqua Group and publication of its first book. Three of the books produced by the group since that time, *Tenders and Contracts*, *Pre-Contract Practice* and *Contract Administration* have long been established as essential reading for students, and standard works on good practice for the building team.

Pre-Contract Practice, now in its seventh edition, describes in detail the procedures leading up to tender. *Contract Administration* (also the seventh edition) takes up procedures from receipt of tenders to placing of the contract, and through the whole post-contract period to settlement of the final account.

The standard forms of building contract are among the most valued tools of the construction industry. The rules and the procedures they lay down generate confidence and contribute to efficiency. They are, of course, subject to periodic revision or indeed may be completely rewritten, as we have seen with the 1980 JCT Standard Form (JCT80) which replaced the 1963 edition. Despite these changes the underlying principles of contract administration remain the same and it is on the rules of good practice that the book focuses. Our third book *Tenders and Contracts* describes in some detail the various other forms of JCT contract.

However, in both *Pre-Contract Practice* and *Contract Administration* we concentrate on the use of JCT80 as a primary form of contract, central as it is to the legal and procedural practices of the building industry and the form on which contract teaching for architects, quantity surveyors and contractors is based.

No matter which form is in use, many clients require alterations to be made to the contract documents. We would like to stress that to avoid confusion and (most important) unintended consequence, such amendments should only be carried out by specialists. It is very easy for the non-specialist to make alterations which fail to achieve the desired aim, or worse, bring about some disastrous result for his client.

In the book we refer to the 'architect' but it will be appreciated that in some circumstances, and certainly when the local authorities edition of the contract is in use, such reference would apply equally to the 'contract administrator'.

At the pre-contract stage everything of consequence must be committed to paper in the form of drawings, schedules, bills of quantities, specification notes and so on. In the course of this book we demonstrate the important part that good paperwork plays in the smooth running of a building contract. It is not a question of creating paper just to fill office files, but rather that clear instructions and good records lead to efficient, economic building and to a fair settlement at the end of the contract free from argument and dispute. It is therefore assumed that tenders have been obtained on the principles set out in The Code of Procedure for Single Stage Selective Tendering published by the National Joint Consultative Committee (NJCC). If so, the contractor will have been provided with comprehensive information about the work itself and all details of the form of contract and will have been allowed adequate time for the preparation of the tenders.

All this effort will be wasted unless the contractor, once appointed, is able to assume responsibility for the site and building work in a smooth and efficient manner. A contract which starts badly, finishes badly. Throughout the contract the reliable performance of the consultants is crucial to the success of the project.

This new edition of *Contract Administration* takes account of the innumerable amendments to the Standard Form of Contract that have been issued by the Joint Contracts Tribunal since the new contract came into being in 1980. The Aqua Group is particularly indebted to Philip Fidler ARICS, FCIArb, of the Bucknall Austin Contract Advisory Service, for his considerable and invaluable contribution in updating Chapters 1 to 8.

Chapter 9, Insolvency (previously titled Liquidation) has been rewritten in the light of recent developments in the law on insolvency. The original chapter and this revision are the work of Kevin J. Greene, solicitor, and we are indebted to him for his important contribution.

We are also grateful to the Royal Institution of Chartered Surveyors, RIBA Publications, the Institute of Clerks of Works of

Great Britain and the National Joint Consultative Committee for Building for permission to reproduce a number of their standard forms.

Finally we thank Brian Bagnall for the contribution of his cartoons. No Aqua Group book would be complete without them.

The Aqua Group comprises:

BRIAN BAGNALL BArch (L'Pool)
TONY BRETT-JONES CBE, FRICS, FCIArb
HELEN DALLAS Dip Arch, RIBA
PETER JOHNSON FRICS, FCIArb
JOHN OAKES FRICS, FCIArb
RICHARD OAKES FRICS
QUENTIN PICKARD BA, RIBA (Chairman)
GEOFFREY POOLE FRIBA ACIArb
GEOFF QUAIFE ARICS
COLIN RICE FRICS
JOHN TOWNSEND FRICS, ACIArb
JOHN WILLCOCK Dip Arch, RIBA
JAMES WILLIAMS Dip Arch (Edin), FRIBA

Chapter 1

The Building Team

The parties to a building contract will be the employer and the contractor. Those appointed by these two will form the 'building team'. The team will usually comprise:

The design team:

- Employer
- Architect
- Quantity surveyor
- Structural consultant
- Services consultants

The clerk of works:

The contractor
- Director, contracts manager and supporting staff
- Site agent (or foreman, described in contract as the person in charge)
- Nominated sub-contractors

"At this first site meeting I'd like to introduce our project manager"

brian bagnall

Of its members the employer, the architect, the quantity surveyor, the contractor, the clerk of works and the nominated sub-contractors are mentioned in the JCT Standard Form of Building Contract (JCT 80). The consulting engineers are not, and their position will depend largely on what form of agreement they have with the employer or the architect. Where they have site inspection duties they should be named in the bills of quantities, but they have no power under the contract and if they wish to issue instructions this must be done through the architect.

The clerk of works is normally appointed by the employer to act, under the direction of the architect, solely as an inspector of the works. He is likely to be an experienced tradesman e.g. carpenter and joiner or bricklayer, and will usually be recommended by the architect. He should be ready to take up his duties about two weeks before the possession date and will be resident on site for the duration of the contract. On larger buildings, particularly those with a high service element, he will be assisted by specialist clerks of works whose appointments will be for varying periods; in some instances resident engineers will fulfil these functions.

It is as well to remember that the building contract is between the employer and the contractor, and although the architect and the quantity surveyor figure in it, and many clauses include the words 'the Architect shall . . . ', these two professional advisors are not parties to the agreement. Therefore if the architect or quantity surveyor fails to carry out any of his duties as defined in the agreement and the contractor considers he has a grievance, his only contractual redress is with the employer.

JCT 80 is exhaustive on the subject of the rights, duties and liabilities of various members of the building team, and each member of the team should be familiar with the contract as a whole, and particularly with those clauses directly concerning his own work. It should be noted that some of the duties are mandatory. References are to the 1980 Standard Form Private Edition with Quantities. For simplicity, clause references have generally been given to the first decimal number. Important rights, duties and liabilities as they concern individual members of the building team are listed below:

The employer

Articles

Name and address
Appointment of architect
Status under statutory tax deduction scheme
Rights, liabilities and procedure in respect of arbitration

Clause

4.1   Right to employ others if contractor does not comply with instructions

5.1   Custody of contract documents (Local Authority edition only)

5.7   Duties in relation to confidential nature of contract documents

12   Right to appoint clerk of works

18.1   Procedure as to partial possession by the employer

19.1   Rights as to assignment of contract

19.3   Rights and procedure as to sub-letting by contractor

20 ⎫
21 ⎭   Rights, duties and liabilities as to injury to or death of persons and damage to property, and insurance relating to this.

22A   Rights when insurance of the works is responsibility of contractor

22B   Duty to insure works as an alternative to 22A

22C   Duty to insure works and existing structures when works are extensions to or alterations in an existing building

22D   Rights to require contractor to insure against delays caused by 'Specified Perils' defined in clause 1.3

23.1   Duty and rights to defer giving contractor possession of site

24.2   Right to deduct damages in respect of non-completion

27.1   Right to determine contractor's employment on contractor's default

27.3   Rights to determine contractor's employment in the event of corruption

27.4   Rights and duties in event of determination of contractor's employment due to contractor's default or bankruptcy

28.2   Rights and duties in event of determination of contractor's employment by contractor due to defaults by employer

28A   Rights and duties in respect of determination of contractor's employment by either party when neither party is at fault

29   Rights and duties in respect of work on site not forming part of contract

30.1   Duty to pay contractor within 14 days of issue of interim certificate
Right to make deductions from monies due

30.4   Right to hold retention in interim payments

30.5   Duty to place retention in separate bank account if required to do so by contractor or nominated sub-contractor (Private edition only)
Duty to inform contractor of any deductions made against retention

31   Duties in connection with statutory tax deduction scheme (where employer is a 'contractor')

32   Rights and duties in event of war

33   Rights as to war damage

34   Rights regarding antiquities found on site

35.13   Right and/or duty to pay nominated sub-contractors direct in event of contractor's default

35.18   Rights if nominated sub-contractor, after early final payment, fails to remedy defects

35.24   Rights in event of determination of nominated sub-contractors' employment

Supplemental Provisions

Rights and duties in regard to VAT

The contractor

Articles

Name and address

Rights, liabilities and procedure in respect of arbitration

Clause

1.4
2.1 } General obligations

2.3 Duties as to discrepancies
4.1 Compliance with architect's instructions
4.2 Right to challenge architect's instructions
4.3 Procedure in connection with verbal instructions
5.3 Duty to provide and update master programme if clause 5.3.1.2. has not been deleted
5.5 Duty to keep documents on site
5.6 Duty to return drawings to architect if asked
5.7 Duties as to confidential nature of contract documents
6.1 Duties as to statutory obligations
6.2 Duty to indemnify employer against statutory charges
7 Duties in setting out
8.1 Duty to comply with standards of materials, goods and workmanship
8.2 Duty to provide proof of compliance and right to be informed within a reasonable time of unsatisfactory work
8.3 Liability in respect of work opened up for inspection or testing and which is shown not to be in accordance with the contract
8.4 Rights and duties in relation to work not in accordance with the contract
9.1 Duty to indemnify employer against royalty and patent claims
10 Duty to keep competent person-in-charge
11 Duty to provide access for architect to job and workshops
12 Duty to afford facilities for clerk of works
13.5 Rules for valuing variations
Duties regarding daywork vouchers
13.6 Right to attend measurement of variations
16.1 Ownership and responsibility for unfixed materials on site
16.2 Ownership and responsibility for unfixed materials off site
17.2
17.3
17.4 } Duties and rights as to making good defects
17.5
18.1 Duties and procedure as to partial possession by the employer
19.1 Rights as to assignment of contract

19.2
19.3 } Rights and procedure as to sub-letting
19.4

20 }
21 } Rights, liabilities, and duties as to injury to or death of persons and damage to property, and duties to insure in respect of these

22A   Duty to insure works

22B }
22C } Duties and rights when insurance of the works is responsibility of employer

22D   Duty, if required, to insure against delays caused by the 'Specified Perils' defined in clause 1.3

23.1   Duty to proceed diligently with the works

24.2   Liability for damages in event of non-completion

25.2   Duty to give notice to architect and sub-contractors of delays; and information to be given

25.3   Duty to prevent delay

26.1   Right to recover loss and expense incurred by matters materially affecting progress of the works

Duties as to notification and information to be provided

26.4   Duties in respect of claims by nominated sub-contractors

27.4   Rights and duties in event of determination of contractor's employment by employer

28.1   Right to determine contractor's employment in the event of employer's defaults and grounds for determination

28.2   Rights and duties in event of determination due to employer's defaults

28A   Rights and duties in respect of determination of contractor's employment by either party when neither party is at fault.

29   Rights and duties in respect of work on site not forming part of contract

30.1   Right to payment under interim certificate within 14 days

30.3   Duties in respect of off-site materials included in interim certificates

30.5   Right to require employer to place retention money in separate bank account (Private edition only)

30.6   Duty to provide documents necessary for adjusting the contract sum

Right to receive copy of computation of adjusted

contract sum
31   Duties in connection with statutory tax deduction scheme (where employer is a 'contractor')
35.2   Right to tender for sub–contract works
35.4   Right to object to nominated sub-contractor
35.5–   } Procedure and duties regarding proposed nomination
35.10   } of sub-contractors by basic method
35.11   } Ditto by alternative method
35.12   }
35.13   Duty to discharge interim payments to nominated sub-contractors
Duty to provide proof of payments to nominated sub-contractors
35.24   Duties in connection with re-nomination after default or determination by nominated sub-contractor
Duty to obtain instruction before determination of nominated sub-contractor's employment
38   }
39   } The Rights and duties in relation to choices available to employer for dealing with fluctuations
40   }

Supplemental Provisions

Rights and duties in regard to VAT

The architect

Articles

Name and address
Duties and procedure in respect of arbitration

Clause

2.3   Duty as to discrepancies between documents
3   Duty to include ascertained amounts in interim certificates
4.2   Duty to justify instructions
4.3   Duty to issue instructions in writing
Procedure in connection with verbal instructions

5.1   Custody of contract documents (Private edition only)
5.2 ⎫
5.3 ⎬   Duties concerning furnishing copies of drawings and
5.4 ⎭   documents
5.6   Right to require return of drawings on completion
5.7   Duties in relation to confidential nature of contract documents
5.8   Procedure for issue of architect's certificates
6.1   Duty to issue instructions in connection with statutory requirements
7   Duties as to setting-out
8.2   Right to require proof of standards of materials etc.
8.3   Rights as to opening up of suspect work
8.4   Rights as to removal of faulty work
8.5   Right to order exclusion of persons from the works
11   Right of access to job and workshops
13.2   Right to issue instruction requiring variation
13.3   Duty to issue instructions regarding provisional sums
16.1   Rights regarding removal of unfixed goods
17.1   Duty to issue certificate of practical completion
17.2 ⎫
17.3 ⎪
17.4 ⎬   Duties as to defects
17.5 ⎭
18.1   Duties concerning partial possession by the employer
19.2 ⎫
19.3 ⎭   Duties and procedure as to sub-letting by contractor
21.1   Right to require evidence of insurance by contractor
21.2   Duties in regard to insurances against damage to property (other than the works) where not caused by contractor's negligence etc
22   Duties and rights in connection with insurance of the works
23.2   Rights regarding postponement of work
24.1   Duty to issue certificates in event of non-completion
25.3   Duty to grant extensions of time and to fix new completion date: information to be given
26.1   Duty to ascertain loss and expense incurred by contractor
26.3   Duty to give details of extension of time granted
26.4   Similar duties in respect of nominated sub-contractors
27.1   Procedure for determining contractor's employment on behalf of employer
27.4   Duties in event of determination of contractor's employ-

ment by employer
30.1   Duty to issue interim certificates
30.3   Discretion to include off-site materials in interim certificate
30.5   Duty to prepare and issue statements of retention in respect of each interim certificate
30.6   Duty to inform contractor and sub-contractors of final valuations of work of nominated sub-contractors
30.7   Duty to issue interim certificate including all final amounts due to nominated sub-contractors
30.8   Duty to issue final certificate and to inform each nominated sub-contractor of date of issue
32   Duties in connection with determination of contractor's employment in event of war
33   Rights concerning war damage
34   Duties relating to antiquities found on site
35.5–35.10   Procedure and duties for nomination of sub-contractor by basic method
35.11–35.12   Ditto by alternative method
35.13   Duties regarding interim payments to nominated sub-contractors
35.14   Duty to operate provisions of sub-contract in dealing with applications for extensions of time
35.15   Duty to certify if nominated sub-contractor fails to complete in time
35.16   Duty to certify practical completion by nominated sub-contractor
35.17–35.18   Duties regarding early final payment to nominated sub-contractor
35.23   Duties when proposed nomination does not proceed
35.24   Duty to re-nominate if nominated sub-contractor defaults or determines sub-contractor's employment
35.25   Duties in connection with determination of nominated sub-contractors' employment
36.2–36.4   Duty, subject to conditions, to issue instructions regarding nominated suppliers

The quantity surveyor

Articles

Name and address

Clause

5.1	Custody of contract documents (Private edition only)
5.7	Duties as to confidential nature of contract documents
13.4	Duty to value variations
13.5	Rules for valuing variations
26.1	Duty to ascertain loss and expense incurred by contractor, if so instructed
26.4	Similar duty in respect of nominated sub-contractors
30.1	Duty to make valuations for interim certificates when required
30.5	Duty to prepare statement of retention in respect of each interim certificate if so instructed
30.6	Duty to prepare statement of final valuations of work of nominated sub-contractors
34	Duties relating to antiquities found on site
38.4 } **39.5**	Duties regarding valuation of fluctuations
40.2	Duty to make valuations for all interim certificates when formula applies for adjusting fluctuations

The person-in-charge

Clause

10	Standing in relation to instructions

The clerk of works

Clause

12	Definition of duties
34	Duties regarding antiquities found on site

Sectional completion supplement

Where this supplement is incorporated in the contract to provide sectional completion by phases, the duties and rights of the employer, contractor, architect and quantity surveyor are clearly defined.

Many of the clauses quoted above will have to be invoked only if the contract is not running smoothly and the primary duty of the building team is to see that it does. It is generally agreed that the origin of contractual disputes is seldom found in actual dishonesty or plain incompetence of any party, but rather in the failure of the architect, contractor, or whoever it may be to put his intentions or thoughts across successfully – in other words a failure of communications.

'Communications' have been the subject of books, conferences and papers and it is a matter with which we cannot hope to deal exhaustively here. Nevertheless we set out below certain golden rules to be observed by all members of the building team in their dealings with each other:

- do not tamper with the standard clauses of the building contract, unless the client insists upon it and even then employ a specialist
- ensure that the contract documents remain in the custody of the architect or quantity surveyor and that all users have certified copies
- use realistic figures in the appendix to the contract
- where alternatives exist or where entries have to be made in the appendix, this information must also be in the bills of quantities at tender stage
- all instructions to the contractor must be channelled through the architect
- all instructions to sub-contractors or suppliers must be channelled through the contractor
- for routine matters such as instructions, site reports, minutes of meetings, and valuations for certificates, use standard forms rather than letters
- circulate to those who need to be kept informed as well as to those who need to act
- be precise and unambiguous
- act promptly

Examples of suggested standard layouts for the more important communications passing between members of the building team are given in later chapters.

Chapter 2

Placing the Contract

The placing of the contract is a relatively simple routine matter but the events which immediately precede it and those which follow immediately afterwards are of great importance.

The receipt of tenders and their examination are discussed in *Pre-Contract Practice*. It may be assumed, therefore, that the quantity surveyor, after examining the bills of quantities of the lowest tenderer, will have reported to the architect and design team, who will in turn have submitted a report to the employer. The employer should be encouraged to make an early decision on that report for it is of the utmost importance to a contractor to know quickly whether or not his tender has been successful.

If no serious errors have been found in the bills of the lowest tenderer the design team's report will normally recommend acceptance of that tender. In making their choice the team will take into account special factors which may be more important than price, such as the contractor's approach to specific problems, his method of working, his programme. Careful consideration may conclude that the lowest tender is not necessarily the best value for money and the team will recommend accordingly. Where the widely accepted practice of selective tendering has been followed, acceptance of other than the lowest tender should be considered only in the most exceptional circumstances.

Probably the employer will accept the team's recommendation and as soon as he does so all contractors who tendered should be notified and sent a list of the tenders received. If priced bills of quantities have been submitted at the same time as the tenders, these should be returned to the unsuccessful contractors unopened.

Errors in bills of quantities

In the course of his examination of the priced bills of quantities the quantity surveyor may have found some errors in pricing or arithmetic, or perhaps in both. If these were of a very minor nature they may be ignored. If, however, they were more serious the contractor should have been advised of them before the report is submitted to the employer.

The invitation to tender will have stated which of the alternatives under section 6 of the NJCC Code of Procedure for Single-stage Selective Tendering 1977 was to apply. Under alternative 1 the contractor will have been given the opportunity of withdrawing if he is not prepared to stand by his tender and under alternative 2 he will have been given the opportunity of correcting genuine errors.

Whichever circumstances apply, errors must be dealt with in the appropriate way in order to put the bills of quantities right for their use as a contract document.

If alternative 1 applies and the contractor has agreed to stand by his tender the errors should be put right, the arithmetic corrected and the summary amended as necessary. An adjustment should then be made at the end of the summary which will leave the final total of the bills equalling the original tender figure. This adjustment will be a lump sum equal to the net amount of the errors and will be added to or deducted from the corrected total of the summary. A note should be added in which

'. . . prepared to stand by his tender . . .'

the amount of the adjustment is expressed as a percentage of the total value of the general contractor's work (i.e. the total of the bills less preliminary items, contingencies and prime cost and provisional sums). Any rates in the bills subsequently used for valuing variations or interim certificates will then be adjusted by this percentage.

If alternative 2 applies, the errors will be corrected in the same way and the summary amended, but no adjustment will be made to restore the total to the original tender figure, the revised summary total becoming the contract sum.

Signing the contract

While the contract documents are being prepared the architect should settle with the contractor the dates for possession of the site and completion of the work, if these have not already been decided. He should make arrangements with the contractor for the initial site meeting and at this stage he should also ensure that the contractor has no valid objection to any nominated sub-contractor or supplier.

Once these matters have been settled the contract documents should be signed. These normally consist of the Articles of Agreement, the drawings showing the extent and nature of the works and the bills of quantities.

It must be borne in mind that in addition to completing the Articles of Agreement at the front of the JCT form of contract, it is also necessary to make a number of deletions or amendments in the text of the conditions. These are as follows:

- Clause **5.3.1.2** Under this clause the contractor is required to provide the architect with copies of his master programme for the execution of the works and subsequently to amend it in the event of an extension of time being granted. If this is not required this clause should be deleted and in the following clause, 5.3.2. the words in parentheses which refer to the master programme should also be deleted.
- Clauses **22A, 22B, 22C** Two of these clauses must be deleted, according to whether the employer or the contractor is responsible for insurance of the works. There is a reminder in a footnote to these clauses that it is sometimes not possible to obtain insurance in the precise terms required by the contract in which case the contract must be amended accordingly.

- Clause **35.13.5.4.4** This clause will require amendment as indicated in the footnote in the rather unlikely event of the contractor being an individual or a company not incorporated under the Companies Acts.
- Clause **41.7** This clause needs amending if the parties do not want the law of the contract to be the law of England and/or do not want the Arbitration Acts 1950 and 1979 to apply to the resolution of disputes.

These deletions and amendments will have been notified to the contractors in the bills of quantities at the time of tendering and no other deletions or amendments should be made without the prior agreement of the contractor. All deletions and amendments must be initialled by the parties at the time the contract documents are signed.

It is also necessary to complete the appendix to the conditions of contract and here again the information to be inserted will have been stated in the bills of quantities. In the event of any item in the appendix having been left for decision until after tenders have been submitted, the matter must be agreed with the contractor before the contract documents are completed.

There are three supplements to JCT 80 which must also be dealt with at this stage. The first of these is the supplemental provisions dealing with value added tax. This is bound in with JCT 80 and must be completed and signed in all cases. The second supplement is appropriate in dealing with sectional completion of the works, which will be discussed later. The third is used if the employer requires the contractor to undertake the design of a portion of the works and operates in a similar fashion to the Form of Building Contract with Contractor's Design (JCT 81).

Returning to the sectional completion supplement, it is important to distinguish between this and clause 18 of JCT 80 – Partial Possession by Employer. The sectional completion supplement enables JCT 80 to be adapted so as to be suitable for use where the works are to be completed by phased sections. It can only be used where tenderers are notified that the employer requires the works to be carried out in phased sections, of which the employer will take possession on practical completion of each section. If the work has not been divided into sections in the tender documents the supplement cannot be used. In such cases, if the employer wishes to take possession of parts of the work during the course of the contract, the provisions of clause 18 will apply but it must be noted that under that clause prior consent of the contractor must be obtained.

If provision has been made for including the sectional completion supplement in the contract documents, the first page of the Articles of Agreement in the form must be amended or replaced by the equivalent page in the supplement. In addition the appendix in JCT80 must be deleted and replaced by the supplement appendix and in accordance with the notes in the supplement.

Prior to preparation of the contract documents, both parties should be consulted as to whether or not the contract is to be under seal. Many parties to contracts prefer to have them executed under seal in order to obtain a twelve year period in which to commence actions for breaches of contract, instead of the six year period applicable to contracts under hand. These limitation periods are provided for in the Limitation Act 1980. In the event of one party sealing the contract and the other signing it, the twelve year limitation appears to apply only against the party who sealed. By virtue of the Finance Act 1985 stamp duty no longer applies to contracts under seal.

The Limitation Act 1980 also deals with limitations for actions in tort, and the Latent Damage Act 1986 amends the 1980 Act by setting time limits for negligence actions in respect of latent damage not involving personal injuries. Although not entirely clear from the 1986 Act itself there are grounds to support the view that it applies to actions in tort and not in contract. The time-limits for commencing actions of the type dealt with by the 1986 Act are:

- 6 years from the actual occurrence of damage
- if later, 3 years from when the damage could reasonably be discovered as set out in the Act.

Both of these, however, are over-ridden by a 15 year long-stop period running from the occurrence of the negligent act or omission causing the damage.

Clause 5 of the conditions of contract requires that the contract documents be held by the architect or quantity surveyor so as to be available at all reasonable times for inspection by the employer or the contractor.

Performance bond or parent company guarantee

A performance bond is a three party agreement between the

employer, the contractor and a surety who agrees to pay a sum of money to the employer, in the event of a default by the contractor. The National Joint Consultative Committee for Building (NJCC) has issued Guidance Note 2: Performance Bonds, which briefly explains the subject and gives an example which is reproduced at the end of this chapter as Example 1.

A parent company guarantee serves a similar purpose to a bond in that it attempts to protect one party to a contract from the effects of defaults by the other party if the latter is a subsidiary company having limited assets. An example is given at the end of this chapter in Example 1A. It is important, where a bond or parent company guarantee is to be provided, that the details should be examined by the design team and possibly the employer's legal advisors before signing the contract, to ensure that it fully covers the requirements set down in the tender documents.

Issue of documents

In accordance with clause 5, immediately after the contract has been signed (and where appropriate, sealed), the architect must furnish the contractor with:

- one copy of the contract documents certified on behalf of the employer
- two further copies of the contract drawings
- two copies of the unpriced bills of quantities

Directly following the execution of the contract, the contractor is to be provided with:

- two copies of any descriptive schedules
- two copies of any other like document necessary for use in carrying out the work
- two copies of further drawings or details to enable the contractor to carry out the works

Providing clause 5.3.1.2. has not been deleted, as soon as possible after the execution of the contract the contractor shall provide the architect with two copies of the master programme.

Other documents which may be necessary for use in carrying out the works from the outset, are as follows:

- party wall agreements

- schedules of condition of adjoining properties
- conditional planning permissions
- building regulation approval including notices to be served during the course of the contract
- estimates from statutory authorities for services
- procedures or estimates for works to be carried out by local authorities such as pavement crossovers, sewer connections

Insurances

The insurances required by the contract are dealt with in clauses 21,22,22A,22B,22C, and 22D.
Clause 21 deals with four aspects:

(1) It reminds the contractor to insure his employees (as must anyone who employs people) under the Employer's Liability (Compulsory Insurance) Act 1969.
(2) It requires the contractor to insure against injury to or death of persons (other than his employees etc.) to the extent of the cover entered in the contract appendix.
(3) It requires the contractor to insure against damage to property (other than the works) to the extent of the cover entered in the contract appendix.
(4) It provides the employer with the option to require the contractor to take out a joint names insurance policy, (ie. the contractor and employer are jointly insured) to cover property adjoining the works against damage caused by the carrying out of the works. This is damage other than that:
  (a) caused by the contractor's negligence etc.
  (b) caused by errors or omissions in the' designing of the works
  (c) which is reasonably forseeable
  (d) arising from war risks or excepted risks i.e. risks which insurance companies exclude from the cover they are prepared to provide
  (e) covered by any insurance under clause 22C.1 taken out by the employer – see later

The extent of any cover will be as provided for in the contract appendix. Clause 22 deals with insurance against damage to the works. The cover is against all risks except for those given as exclusions in clause 22.2. Clause 22A provides for the contractor to take out a joint names policy where the works comprise a new building. Clause 22B is used instead if the works comprise a new building and the employer has elected to take out the joint names insurance policy. Clause 22C is used instead of 22A or B if the works are an extension to or an alteration in an existing building. In this case the employer is required to take out two joint names insurance policies:

- to insure the works against all risks as previously described
- to insure the existing building other than the works (referred to in clause 22C.1 as existing structures and contents) against damage caused by the specified perils defined in clause 1.3.

The amount of cover is the full cost of re-instatement including professional fees and contents as appropriate. In respect of the contractor's insurance under clause 22A it is common for the contractor to maintain an annually renewable insurance policy which provides cover of no less than that required by clause 22A. Clause 22A.3 sets out deemed to satisfy provisions in respect of such annual policies.

Construction insurance is an extremely complex subject. It is further complicated by the fact that the policies available in the market do not have standardised wording. Advising on insurances is therefore very difficult and for all but extremely simple situations the client should be recommended to consult an insurance broker.

Clause 22D provides the employer with the option to require the contractor to insure against delays to completion of the works caused by any of the specified perils defined in clause 1.3. The amount of cover will be at the rate stated in the contract appendix for liquidated and ascertained damages for the period given in the appendix. This type of insurance is currently expensive and, if required, is usually in respect of delays of no more than ten weeks.

The architect must therefore ensure, before the commencement of works on site, that all necessary insurances are in place. He should also satisfy himself that premium renewals falling due during the works on site are properly paid. Also in the event of the employer taking possession of parts or sections before completion of the whole works, or in the event of the employer having items of plant or other contents he has paid for direct, stored on site or in the works before completion, the architect should ensure that, where necessary, policies are amended accordingly.

Analysis of tender

Finally, during the process of the placing of the contract the quantity surveyor will carry out any analysis of the tender that the employer or the architect may require, at the same time updating his cost plan, thus establishing a proper basis for cost control during the execution of the work.

EXAMPLE 1: PERFORMANCE BOND

PERFORMANCE BOND

THIS BOND is made the _____ day of _____ 19 _____

BETWEEN _____

of _____
(hereinafter called 'the Contractor') of the first part and

_____   _____

of _____
(hereinafter called 'the Surety') of the second part and

_____

of _____
(hereinafter called 'the Employer') of the third part.

1   By a contract dated _____ made between the Contractor
    and the Employer (hereinafter called 'the Contract') the Contractor has agreed to carry
    out the Works specified in the Contract for the sum of £ _____ _____

2   The Contractor and the Surety are hereby jointly and severally bound to the Employer
    in the sum of £ _____ (not exceeding ten per cent of the original
    contract sum) which sum shall be reduced by an amount equal to ten per cent of the
    value of any part or parts of the Works taken into the possession of the Employer under
    the provisions of the contract * *or of any section of the Works upon the Architect/*
    Supervising Officer certifying practical completion of that section; provided that if the
    Contractor shall, subject to Clause 5 hereof, duly perform and observe all the terms,
    conditions, stipulations and provisions contained or referred to in the Contract which
    are to be performed or observed by the Contractor or if on default by the Contractor the
    Surety shall, subject to Clause 3 hereof, satisfy and discharge the damage sustained by
    the Employer thereby up to the amount of this Bond then this agreement shall be of no
    effect.

3   If the Contractor has failed to carry out the obligations referred to in Clause 2 hereof,
    then written notice requiring the Contractor to remedy his failure, where possible, shall
    be given, and if the Contractor fails so to do or repeats his default the Employer shall
    be entitled to call upon the Surety in accordance with Clause 2.

4   Any variations required under the Contract shall not in any way release the Surety
    from its obligations under this agreement.

5   The Contractor and the Surety shall be released from their respective liabilities under
    this agreement upon the date of Practical Completion of the Works as certified by the
    Architect/Supervising Officer appointed under the Contract.

The Common seal of the Contractor
was hereunto affixed in the presence of:

_____

The Common seal of the Surety
was hereunto affixed in the presence of:

_____

It should be noted that this is an example not a model form.
The wording in italics covers sectional completion.

EXAMPLE 1A: PARENT COMPANY GUARANTEE

THIS DEED is made the . . . . , ... day

of . . . . 19 . . . . BETWEEN . . . . whose

registered office is at . . . . (hereinafter

called "the Guarantor") of the one part and . . . .

. . . . whose registered office is at

. . . .

(hereinafter called "the Employer") of the other part

WHEREAS

(1) This Agreement is supplemental to a contract (hereinafter

called "the Contract") dated the . . . . day of . . . . 19 . . . .

and made between the Employer of the one part and . . . .

whose registered office is at . . . .

(hereinafter called "the Contractors") of the other part whereby the Contractors agreed and undertook to

carry out the following works . . . .

. . . .

. . . .

(2) The Guarantor has agreed the due performance of the contract in manner hereinafter appearing

NOW THIS DEED WITNESSETH as follows:–

1. The Guarantor hereby covenants with the Employer that the Contractor will duly perform the obligations on the part of the Contractors contained in the Contract and that if the Contractors shall in any respect fail to execute the Contract or commit any breach of any of their obligations thereunder then the Guarantor will be responsible for and will indemnify and keep indemnified the Employer from and against all losses damages costs and expenses which may be suffered or incurred by it by reason of or arising directly or indirectly out of any default on the part of the Contractors in performing and observing the obligations on their part contained in the Contract.

In WITNESS whereof the Guarantor
have caused their Common Seal to
be hereunto affixed this day and
year first above written

THE COMMON SEAL OF

EXAMPLE 2 *(page 1 only of six page document included)*

Incorporating Amendments 1: 1987 and 2: 1988   Agreement NSC/2

JCT Standard Form of Employer/Nominated Sub-Contractor Agreement

Agreement between a Sub-Contractor to be nominated for Sub-Contract Work in accordance with clauses 35·6 to 35·10 of the Standard Form of Building Contract for a main contract and the Employer referred to in the main contract.

Main Contract Works: _____

Location: _____

Sub-Contract Works: _____

This Agreement

The date to be inserted here must be the date when the Tender NSC/1 is signed as 'approved' by the Architect/the Contract Administrator on behalf of the Employer

is made the _____ day of _____ 19 _____

between _____

of or whose registered office is situated at _____

_____

(hereinafter called 'the Employer') and

_____

of or whose registered office is situated at _____

_____

(hereinafter called 'the Sub-Contractor')

Whereas

First   the Sub-Contractor has submitted a tender on Tender NSC/1 (hereinafter called 'the Tender') on the terms and conditions in that Tender to carry out Works (referred to above and hereinafter called 'the Sub-Contract Works') as part of the Main Contract Works referred to above to be or being carried out on the terms and conditions relating thereto referred to in Schedule 1 of the Tender (hereinafter called 'the Main Contract');

Second   the Employer has appointed

_____

to be the Architect/the Contract Administrator for the purposes of the Main Contract and this Agreement (hereinafter called 'the Architect/the Contract Administrator' which expression as used in this Agreement shall include his successors validly appointed under the Main Contract or otherwise before the Main Contract is operative);

Third   the Architect/the Contract Administrator on behalf of the Employer has approved the Tender and intends that after agreement between the Contractor and Sub-Contractor on the Particular Conditions in Schedule 2 thereof an instruction on Nomination NSC/3 shall be issued to the Contractor for the Main Contract (hereinafter called 'the Main Contractor') nominating the Sub-Contractor to carry out and complete the Sub-Contract Works on the terms and conditions of the Tender;

Fourth   nothing contained in this Agreement nor anything contained in the Tender is intended to render the Architect/the Contract Administrator in any way liable to the Sub-Contractor in relation to matters in the said Agreement and Tender.

Chapter 3

Progress and Site Meetings

Initial site or briefing meeting

As soon as practicable after the contract has been placed, the building team should meet. Although this initial meeting may take place on the site, it will more probably take place in either the architect's or contractor's office.

The manner in which this first meeting is conducted will greatly influence the success of the programme and succinct clear direction from the chairman will be a strong inducement to a similar response from the others. Since at this stage the person having the most complete picture of the job is likely to be the architect, it seems logical that he should take the chair. He will have discussed the arrangements for this meeting with the contractor beforehand, and it is suggested that the representatives of the following should attend, as appropriate:

- employer
- architect
- quantity surveyor
- structural consultant
- services consultant
- contractor
- principal nominated sub-contractors
- principal nominated suppliers
- clerk of works
- project managers

It is suggested that the agenda for this meeting should include the following matters:

- introduction of those attending
- factors affecting the carrying out of the work
- programme
- sub-contracts and employer/sub-contractor agreements
- lines of communication

- insurances (see chapter 2)
- procedure to be followed at subsequent meetings

Introduction

The introduction of those attending needs no elaboration, though it is more than just a formality as it establishes an initial contact between individuals who must work together in harmony if the contract is to run smoothly.

Factors affecting the carrying out of the works

These would normally be described fully in the contract documents, but may require emphasis and clarification at this initial meeting.
The factors may include:

- access to site
- space availability
- restrictions such as hours of work and noise
- building lines
- buried services
- site investigation
- protection of the works, unfixed materials, adjoining buildings, work people and the general public
- Health and Safety at Work Act 1974

Programme

The contractor should attend the initial site meeting with an outline programme for the work prepared in advance, the necessary basic information regarding delivery dates and construction times having been obtained from the principal nominated sub-contractors and suppliers. It is helpful to have this programme circulated to those attending before the meeting when it can then be properly considered and adjusted as necessary. Following this the contractor can then prepare the final programme and circulate it to all concerned.

The opportunity should be taken to stress the importance of adhering to dates once these have been agreed by the contractor and the nominated sub-contractors and suppliers. This applies also to dates agreed for the issue of architect's or consultant's drawings, where these have not already been prepared in the pre-contract stage. It is not uncommon for the contractor to indicate on his programme the latest dates by which he requires drawings, instructions for placing orders, schedules and other information from the architect, and the latter must indicate at this stage whether he considers the proposed dates reasonable.

The master programme is not a contract document and clause 5.3.2 makes it clear that nothing contained in the programme can impose any obligation on the contractor beyond the obligations imposed by the contract documents as such.

If clause 5.3.1.2. of JCT 80 has not been deleted it provides that as soon as possible after the effecting of the contract the contractor shall provide the architect with two copies of his master programme for the execution of the works. The clause should be deleted if a master programme is not required by the architect or if the contractor has not agreed to provide such a programme. Experience has shown that a master programme is an essential tool of management both for the contractor and for the architect. Only in exceptional circumstances would a master programme not be required; indeed there is much to be said for contractors tendering being required to submit the master programme with their tenders.

In addition to providing the master programme at the start of the contract the contractor is also required to update it within four days of any decision by the architect which may create a new completion date for the contract.

The judgment in *Glenlion Construction Ltd* v *The Guinness Trust* decided that:

(1) The contractor is entitled to finish the works before the completion date stated in the contract appendix (as extended by the provisions of clause 25).
(2) The contractor is not entitled to expect the design team to provide design information so that he can finish the works before the completion date stated in the contract appendix (as extended by the provisions of clause 25).

At first sight these two statements might appear ambiguous; but further thought will clarify that they are not because it may not be feasible for the employer and design team to provide design information quicker than would be necessary to complete the works by the contract completion date. If they can and all parties agree, then the solution might be for the employer and contractor to amend the contract to provide a mutually agreed shorter period for completion of the works.

Sub-contracts

Many problems on building contracts arise from delay in the issue of instructions by the architect regarding nominated sub-contractors and suppliers and from difficulties in liaison between contractor and sub-contractors. It is advisable at this initial meeting, therefore, to clarify the position regarding all work covered by provisional and prime cost sums in the bills of quantities.

It is essential for the smooth running of the contract that all nominations are made in adequate time for the work concerned to be phased into the contractor's programme without causing disruption. It should be made clear to the contractor that, once nominated, these sub-contractors and suppliers are his responsibility contractually. It is important that the employer, his design team and the contractor allow sufficient time to carry out their various duties regarding the complex and often time-consuming procedures for the selection and appointment of nominated sub-contractors and suppliers.

The importance of proper sub-contract documentation should be stressed. This subject is dealt with in more detail in *Pre-Contract Practice*.

In the past, when nominated sub-contractors or suppliers defaulted and they had been responsible for some aspects of the design of their work and/or for selection of materials and other such matters, it was usually impossible for the employer to obtain redress as the contractor had no such responsibilities under the main contract. In order to overcome this problem it is sensible to arrange direct agreements, such as that in Example 2, between the employer and nominated sub-contractors or suppliers. A recent court judgment now commends the addition of wording that 'reasonable skill and care shall be used in carrying out the work'.

Lines of communication

It is important at the initial site meeting that the procedure regarding architect's instructions should be made clear to all concerned. The matter is covered by clause 4 of the conditions of

'. . . essential for smooth running of contract . . .'

contract and is dealt with in Chapter 5 of this book. Additional points which should be stressed at the initial site meeting will be found in the golden rules of communications at the end of Chapter 1.

Subsequent site meetings

This refers to formal site meetings and should not be confused with the architect's site visits and the numerous meetings between the contractor and others, which may be necessary during the progress of the work. The frequency of site meetings will vary with the size and complexity of the contract, and according to the particular stage of the job and any difficulties encountered. It is unlikely that these meetings at any stage would be at intervals of more than four weeks, and especially in the early stages they may well be held more frequently.

When considering the procedure to be followed, the following points should be borne in mind:

(1)   *Notices to attend*:
      The architect should notify the main contractor, the quantity surveyor and consultants of the dates and times of site meetings, asking them to attend if their presence is required. It should be the responsibility of the contractor to call all sub-contractors' and suppliers' representatives whom he or the architect would like to be present.

(2)   *Agenda*:
      The content of the agenda should be agreed before each meeting by the architect and the contractor. A standard form of agenda is useful, as a model and as an *aide mémoire*. A typical agenda is shown in Example 3.

(3)   *Minutes*:
      Minutes should be impartially drafted and given the correct emphasis; they should be concise and should record decisions reached and action required. Example 4 shows typical site meeting minutes.

The requirements of the employer and/or architect as to the nature, frequency, and procedures for, site meetings should be given in the main contract tender documents. On many large contracts it may be more convenient to divide the site meetings into two parts: the first, under the chairmanship of the contractor, will concern itself with the method of carrying out the work. It would be attended solely by the contractor's representatives and those of sub-contractors and suppliers. It will plan and organise the work on site. The second meeting will monitor progress and performance. It will be chaired by the architect and attended by the design team, the contractor, and such sub-contractors as are requested to attend. The object of this meeting would be to give the contractor the opportunity to raise questions on the drawings, specifications, schedules and instructions issued; to request additional information and to report progress. Consideration of the value of people's time should be shown at all meetings; care should be taken not to call to meetings persons whose presence is not really necessary. It is important to realise that a badly run site meeting can be a serious waste of time to all, but if properly handled it can be a big saver of time. Furthermore, there is little doubt that these periodic meetings keep people up to the mark and impart a sense of urgency which is difficult in day-to-day correspondence.

The contractor should not regard the formal meetings as relieving him of the obligation to manage the job efficiently.

EXAMPLE 3: TYPICAL SITE MEETING AGENDA

Project: Shops & Offices, Newbridge St, Borchester
Project ref: 456
Agenda for site meeting
Date: 7 October 1989 at 10.00 am

1.0	Apologies
2.0	Agree minutes of last meeting

3.0	Contractor's report:
	    General report
	    Sub-contractors' meeting report
	    Progress and causes of delays and claims arising
	    Information received since last meeting
	    Information and drawings required
	    Architect's Instructions required

4.0	Clerk of Works' report
	    Site matters
	    Quality control
	    Lost time

5.0	Consultant's reports
	    Architect
	    Structural Engineer
	    Services Engineer

6.0	Quantity Surveyor's report
7.0	Communications and procedure

8.0	Contract completion date
	    Assess likely delays on contract
	    Review factors from previous meeting
	    List factors for review at next meeting
	    Record completion date (as revised)

9.0	Any other business
10.0	Date, time and place of next site meeting Date & time of next official visit

Distribution:

Client
Main Contractor
QS
Struct Engr
Services Engr
Clerk of Works
File

EXAMPLE 4: TYPICAL SITE MEETING MINUTES

Project title:	Shops & Offices Newbridge St., Borchester	Reed & Seymore Architects 12 The Broadway Borchester BC4 2NW
Project ref:	456	
Meeting title:	4th site meeting	

Date:	7 October 1989
Location:	Site

Those present:

S Gilbert	Client (Aqua Products Ltd)
A Morley	Main Contractor (Leavesden Barnes & Co Ltd)
G MacKay	" "
B Hunt	QS (Fussedon Knowles & Partners)
I Tegan	Struct. Engrs. (GFP & Partners)
I Hills	Architect (Reed & Seymore)
H A Hemming	Clerk of Works
F Adams	Services Engrs. (Black and Associates)

Item		Action
1.0	Apologies None	
2.0	Minutes of last meeting Agreed as correct	
3.0	Contractor's Report	
3.1	Progress is generally satisfactory	
3.2	Still one week behind programme due to late delivery of bricks	
3.3	AI 5 received and actioned	
3.4	Details of ironmongery revisions required in next two weeks	Arch
4.0	Clerk of Works' Report	
4.1	Concern expressed about poor stacking of bricks	MC
	(continue accordingly through agenda)	
9.0	Any other business None	
10.0	Date of next meeting 4 November 1989, 10.00 am on site	All

Distribution:

2	Client	1	Services Engr
3	Main Contractor	1	Clerk of Works
1	QS	1	File
1	Struct Engr		

Chapter 4

Site Duties

The object of site inspection is primarily to ensure that the employer's requirements as expressed in the contract documents are correctly interpreted and that the problems which are bound to arise on even the smallest jobs are satisfactorily resolved. Supervision is the responsibility of the main contractor; he is bound by the conditions of his contract to complete the work in a certain time and to a specified standard. The architect, under the terms of his appointment, is required to 'visit the site at intervals appropriate to the stage of construction to inspect the progress and quality of the works and to determine that they are being executed generally in accordance with the contract documents' (RIBA *Architect's Appointment*, clause 3.10). He is not, however, required to make frequent or constant inspections.

'possibly a resident architect'

The nature and extent of inspection arrangements will depend on the size and complexity of the works. On a small contract for example, the builder may be able to rely on a competent general foreman, while on larger contracts a number of foremen and assistants may be required working under a site agent, if the job is to be properly organised. Similarly the architect on a small contract may be able adequately to undertake his normal duties by periodic visits, whereas one or more clerks of works may be needed and possibly a resident architect, on large and complex buildings. In this chapter we are assuming that there will be a clerk of works acting, as the contract provides, 'as inspector on behalf of the employer under the directions of the architect'.

We are concerned with inspection by the architect and this can be considered under the following headings:

- formal site meetings
- routine site inspections
- records and reports
- samples and testing

Formal site meetings have been dealt with in the preceding chapter. The remaining headings, however, do call for some further comment.

Routine site inspections

The architect will normally have more time to inspect the work when making routine inspections than on those occasions when a formal site meeting is held. The frequency of these inspections will obviously depend on the size and complexity of the work and the speed of the progress being made.

As a matter of courtesy the architect should make his presence on the site known to the clerk of works and the site agent who will normally accompany him around the job. The architect should never give instructions to a workman direct. The main contractor, through his site agent or general foreman, is the only one with authority under the contract to act upon architect's instructions. All instructions should be confirmed in writing after the visit as set out in chapter 5.

The Building Act 1984 consolidates the building control provisions of a number of earlier Acts which imposed certain legal responsibilities upon the contractor and employer. Where the possession of the site is vested in the contractor the primary liability for health and safety rests with him, and the architect should ensure that these responsibilities are met. Where the site is in the possession of the employer, or jointly with the contractor, both the employer and contractor are responsible for the health and safety of the workforce and other persons on site. The contractor, as the expert, is responsible for the plant. Recommendations contained in BS 5306: Part 0: 1986 (British Standard for fire extinguishing installations and equipment on premises) should also be observed, a point which might have been covered in the bills of quantities.

When making a site inspection it is easy to be distracted and to overlook items which require attention. It is a good plan therefore, to list in advance particular points to be looked at and any special reason for doing so. For this purpose a standard check-list is helpful as it serves as a useful *aide mémoire*. The contents of such a list must be of a rather general nature and it can then be amplified or adapted for each job according to the type of work and form of construction involved. A typical standard check-list is given in Example 5.

Records and reports

Where a clerk of works is employed, the architect should receive weekly reports from the site, covering:

- number of men employed in the various trades
- state of the weather and particulars of time lost due to adverse weather conditions
- principal deliveries of materials and particulars of any shortages
- plant on the site
- particulars of any drawings or other information required
- visitors to the site
- general progress in relation to the programme
- any other matters affecting the smooth running of the contract

A standard form of report is available from the Institute of Clerks of Works (as shown in Example 6), or may be provided by the architect.

The weekly report is valuable in keeping the architect fully informed of day-to-day progress, and also constitutes a useful record for reference if disputes arise at a later date. The weekly report should be compiled from the diary of the clerk of works. This diary should be provided by the architect at the commencement of the job and should be returned to him on completion. In it should be recorded daily all matters affecting the contract.

A copy of the builder's programme should be kept in the office of the clerk of works, and every week the actual progress should be checked against this programme.

It is the job of the clerk of works to keep records of any departures from the production information which may be found necessary in order that, on completion, the architect has all the information necessary to enable him to issue to the employer an accurate set of drawings of the finished building. These records are particularly important where the work is to be concealed, as for example foundations, the depth of which may vary from that shown on the original production information.

Progress photographs of the work also form valuable records if taken regularly. This is probably best arranged in conjunction with the contractor so that both he and the architect can have the benefit of them.

Samples and testing

The architect may call for samples of various components and materials required in the building to be submitted for approval in order that he can satisfy himself of their construction or quality and that they meet the requirements of the client and, where applicable, the local planning authorities.

Some of the items, of which samples would normally be required, are:

- External materials, e.g. facing bricks, artificial or natural stone, precast concrete, marble, terrazzo, slates or roofing tiles.
- Internal finishes, e.g. joinery, mouldings, timber, wall or flooring tiles, other decorative finishes
- Services, e.g. plumbing components, sanitary goods, ironmongery, electrical fittings.

In addition to samples of individual components or materials, the architect will often require sample panels prepared on the site to enable him to judge the effect of the materials in the positions in which they will be used. Panels of facing bricks are an example of this.

It is quite normal to require laboratory tests of basic materials such as concrete. The crushing strength of bricks may also be tested in a laboratory, although, unless the design requirements are stringent, a certificate from the manufacturers giving their characteristics may suffice. The testing of concrete should be carried out on a regular basis and cubes should be cast from each main batch of concrete, carefully labelled and identified. The tests themselves should be carried out by an approved laboratory and test reports submitted by the contractor to the architect or structural engineer.

British Standard Specifications, Agrément Certificates and Codes of Practice are specified for many building materials, components and processes, and in carrying out tests it is essential to refer to the appropriate standard or code to ensure that the requirements are complied with. In addition the British Standards Institution lays down acceptable tolerances for manufactured goods, and copies of the relevant standards and codes should always be kept by the clerk of works on the site.

EXAMPLE 5: STANDARD CHECK-LIST FOR SITE INSPECTIONS

This list should be amplified or adapted according to the nature of the project and may serve architect or engineer. Some of the items should be inspected jointly with other consultants.

(1)   General

In all cases check that the work complies with the latest drawings and specification, with the latest requirements of the statutory undertakers and with the building regulations. Ensure that all information is complete.

(2)   Preliminary works

- Scaffolding and the Building Act 1984
- Siting of workmen's canteens and builder's offices, etc.
- Removal of top soil and location of spoil heaps
- Perimeter fencing or hoardings
- Provision for protection of rights of way
- Protection of trees and other special site features
- Party wall agreements and protection of adjoining property
- Protection of materials on site
- Site security generally
- Suitability and siting of clerk of works' or site architect's office
- Agree bench mark or level pegs
- Agree setting out

(3)   Demolition

- Extent
- Adequacy of shoring
- Burning on site
- Preservation of certain materials and special items

(4)   Excavation and foundations

- Width of trenches
- Depths of excavations
- Nature of ground in relation to trial hole report
- Stability and timbering of excavations

- Pumping arrangements
- Risk to adjoining property or general public
- Quality of concrete and thickness of beds
- Suitability of hardcore (freedom from rubbish)
- Quality of sand and ballast (freedom from loam and correct grading)
- Damp-proof membranes or asphalt tanking
- Ducts, drains or services under building
- Correct placing of reinforcement, including diameter, bending and spacing of bars
- Consolidation of backfilling, particularly suitability of material used for backfilling
- Ascertain depths of piles and driving conditions in case of piled foundations
- Local authority inspections

(5) Drainage

- Depths of inverts and gradients of falls
- Timbering of trenches
- Thickness of concrete bed and jointing of pipes
- Quality of bricks for manholes and rendering thereto
- Testing of drains and manholes (water test), etc.
- Local Authority inspections

(6) Brickwork, blockwork and concrete masonry

- Approve sample panels of facings and fairface work
- Quality and colour of mortar and pointing
- Test report on crushing strength where necessary
- BS certificates on load-bearing blocks
- Silt test on sand
- Position and type of wall ties; inspect cavities to external walls to ensure cavities clean above DPCs and wall ties clean
- Correct setting-out and maintenance of regular vertical and horizontal joints
- Solid bedding to horizontal and vertical joints
- Type and quality of DPCs and correct placing
- Correct setting-out and fixings for door frames, windows, etc.
- Bedding and levels of lintels over openings
- Position of fixing blocks, etc.
- Expansion joints

(7) *In-situ concrete*

- Setting out and stability of shuttering
- Correct shuttering to achieve type of finish specified
- Setting out of reinforcement, fixings, holes and water bars
- Mix and correct procedure of taking test cubes
- Curing of concrete and striking of shuttering

(8) *Precast concrete*

- Size and shape of units
- Finish
- Position of fixings, holes, etc.
- Damage in transit and erection
- Neoprene strips, water bars, etc.

(9) *Carpentry and joinery*

- Freedom from loose knots, shakes, sapwood, insect attack, etc.
- Dimensions within permissible tolerances
- Application of timber preservatives and primers
- Storage and stacking and protection from weather
- Jointing, bolting, spiking and notching of carpenter's timber
- Spacing of floor and ceiling joists and position of trimmers
- Spacing of battens, position of noggings
- Weather throatings and hardwood cills to doors and windows, etc.
- Jointing, machining and finish of manufactured joinery

(10) *Roofing*

- Pitch of roof
- Spacing of rafters and tile battens
- Approval of underfelt
- Approval of roofing tiles and nailing as specified
- Pointing to verges
- Bedding of ridges, etc.
- Falls to outlets for flat roofs
- Thickness of insulation under finish
- Chasing of walls and parapets to tuck in asphalt
- Pointing above skirtings and flashings

- Fixing of drips to external edges and finish of asphalt thereto
- Weight of lead to lead roofs and correct formation of flashings, etc.

(11) *Cladding*

- Vapour barriers and insulation
- Regularity of grounds for sheet materials
- Location and quality of fixings (cleats, bolts, nails, etc.)
- Laps, tolerances, and positions of joints
- Setting-out and jointing of mullions and rails
- Glass type, thickness, sizes, tolerances, spacers and fixing
- Glazing compounds
- Handling and protection of panel materials (metal, glass and GRP)
- Specification and application of mastic
- Flashings, edge trims and weather drips
- Entry and egress of moisture
- Prevention of electrolytic action
- Specification of mortar bedding and pointing
- Location and type of movement joints (brick and tile claddings)

(12) *Steelwork*

- Sizes of steel
- Rivets and welding
- Position of steels
- Plumbing, squaring and levelling of steel frame
- Priming and protection

(13) *Metalwork*

- Sizing and spacing of members
- Galvanising or rustproofing, if specified
- Connections between units to be similarly protected
- Stability of supports, including caulking or plugging
- Isolation from corrosive materials, etc.

(14) *Plumbing and sanitary goods*

- Inspect sanitary goods to ensure freedom from cracks or deformities
- Protection against frost

- Location and fixing of stack pipes
- Falls to waste branches
- Use of traps
- Jointing of pipes
- Smoke and/or water tests
- Location and accessibility of valves, stop-cocks
- Drain-down cocks at lowest point
- Access to traps and rodding eyes
- Local authority inspections

(15) *Heating, hot water and ventilation installations*

- Type of boiler, calorifier, fans, etc, as specified
- Types of pipes
- Position and type of stop valves
- Position of pipe runs and ventilation trunking
- Insulation of pipe work
- Identification and labelling of pipes, valves, etc. and directions of flow

(16) *Electrical installation*

- Approval of components
- Switches, fuses, cables, etc., as specified
- Position of heating, lighting and switch points as compared with drawings
- Earthing of installation
- Lightning conductor installation
- Runs of cables and quality of connections, etc.
- Labelling and identification of switchgear, etc.

(17) *Specialist installations*

- Specialist drawings on site
- Drawings of builders work in connection with installation
- Special statutory regulations and controls
- Unforeseen changes in design of special equipment
- Power and plant to be provided during installation
- Access for equipment and provision of adequate working space
- Temporary support, loading on structure, lifting tackle, etc.
- Provision of services
- Provision of correct temperature, humidity and ventilation during installation

- Attendance on site and correct sequence of work
- Accuracy, size and position of holes, levels, plumb lines, and tolerances generally
- Correct type and setting-out of fixings
- Junctions with surrounding finishes
- Safety – operatives in confined spaces, contractor's workforce and the public. (Volatile and noxious gases, radiation, noise, etc.)

(18) *Paving and floor tiling*

- Approve materials
- Quality of screeds to receive flooring
- Junctions of differing floor finishes
- Regularity of finish
- Falls to gulleys, etc.
- Skirtings and coves
- Non-slip surfaces, if specified, etc.

(19) *Plastering*

- Storage of material
- Correct mix
- Preparation of surface
- True surfaces and arrises (including angle and casing heads)
- Fixing of laths or plaster board
- Filling and scrimming of joints in plaster board
- Bonding plaster on concrete or adequate hacking
- Wall tiling, regularity of joints
- External angles

(20) *Suspended ceilings*

- Type of suspension and tile
- Height of ceiling and correct setting out
- Position of light fittings, ventilation grilles, etc.
- Access panels
- Finish for curtains, blinds, etc.

(21) *Glazing*

- Quality of glass and freedom from defects

- Correct type and/or thickness
- Size of sheets to allow fractional movement in frames
- Depth of rebates
- Glazing compound, fixing of glazing beads, etc.

(22)   *Painting and decorating*

- Approval of material being used
- Storage of material safe from frost
- Preparation of surfaces and freedom from damp
- Specified coats
- Take sample from painter's can for test to check against over-thinning
- Inspect surfaces partly concealed, viz insides of eaves, gutters
- Finished work – freedom from runs, brush marks, etc.
- Opacity of finish, etc.

(23)   *Cleaning down and handing over*

- Windows cleaned and floors scrubbed
- Floors free from paint spots
- Sanitary goods washed and flushed
- Painted surfaces immaculate
- Doors correctly fitted, windows not binding or rattling
- Ironmongery complete and locks and latches operating correctly
- Correct number of keys
- Floor or overhead springs correctly adjusted
- Connection of services, provision of meters, etc.
- Commissioning of all mechanical engineering plant, balancing of air-conditioning, etc.
- Plant maintenance manuals, plant room service diagrams, etc.
- Operation of security, communication and fire protection systems

EXAMPLE 6

WORKS PROGRESS REPORT (No. 16)

ARCHITECTS ...... Reed & Seymore ......
NAME OF JOB ...... Shops & offices ......
ADDRESS ...... Newbridge St. Borchester ......
MAIN CONTRACTOR ...... Leavesden Barnes ......

Week/Month ending ...... 8 October 1989 ......
Contract No. ...... 456 ......
Date work started ...... 26 June 1989 ......
Contract completion date ...... 13 July 1990 ......
Forecast completion date ...... 20 July 1990 ......
Progress + or − to programme ...... −1 ...... weeks

LABOUR REPORT

* 'A'—Number on Site on day of Return.   'B'—Estimated Labour Shortage.

	A	B		A	B		A	B		A	B
Building Labour ...			B/Fd. ...	11	1	B/Fd. ...	14	1	Mech. & Eng. Labour		
Agent	1		Glaziers						Electricians		
Foremen and Gangers	1		Drain Layers ...		. .				Electricians' Mates ...		
Bricklayers	4	1	Labourers	2					Fitters ·		
Carpenters	1		Apprentices	1					Fitters' Mates		
Masons			Asphalters						Laggers		
Barbenders	4								Lift Engineers		
Steel Erectors						Total Building Labour	14	1			
Scaffolders						Office Staff	2				
Plumbers						Welfare					
Plasterers						Watchman	1				
Painters											
C/Fd. ...	11	1	C/Fd. ...	14	1	Total ...	17	1	Total ...		

* Completed Daily Labour Returns Enclosed (if required).

Total Labour on Site   17

DAY	Weather Report	Time lost man days	Difficulties including shortages which may cause delays and action taken.
Mon.	Dry & overcast		
Tues.	Dry & overcast		
Wed.	Light rain		
Thurs.	Light rain am, heavy rain pm	½	
Fri.	Intermittent rain		
Sat.	−		
Sun.	−		
Total Man Days		½	

Total Man Days lost to date   3   ~~* See Continuation Sheet~~ (* Delete if not used.)

DRAWINGS AND INFORMATION REQUIRED	VISITORS
Ironmongery revisions	Building Inspector, project architect, quantity surveyor.

ATTACHMENTS TO REPORT. IF NONE, WRITE *NONE*.	FOR USE BY ARCHITECT'S OFFICE
NONE	

Signature of Clerk of Works   Harry Hemmings
Telephone No.   Date 9 October '89

General Report see over

EXAMPLE 6 CONT.

	%		%			%		%
Preliminaries		Patent Roofing					Gas Mains	
Excavations	100	Plumbing					Electric Mains	
Concrete Foundations ...	90	Carpentry					Water Mains	
Piling Sheet		Joinery					Hot Water Service ...	
Piling Foundations ...		Window Fixing					Electrical Service ...	
Asphalt Tanking ...		Plastering					Heating	
Asphalt Roofing ...		Metalwork					Ventilation	
Brickwork External ...	✳	Mosaic and Terrazzo ...					Lifts	
Brickwork Internal ...		Decorating						
Drainage	30							
Paving and Flooring ...								
Masonry								
Slater and Tiler ...								

* STAR ITEMS BEHIND PROGRAMME * and add further comment in General Report

GENERAL PROGRESS. Summary of work proceeding, etc.

One to 1½ weeks behind programme due to late delivery of bricks. Bwk below DPC held up as a result.

Bricks now on site but poorly stacked – Agent informed that height of stacks should be reduced, but site is getting congested.

Rainy weather may hold up footings bwk next week.

Drainage excavation is nearly complete.

Progress is generally satisfactory apart from bricks.

Building Inspector satisfied with depth of footings excavation.

* Omit if not required

→ Continued to Sheet No. II

Clerk of Works H.A. Hemmings

(C) 1964   The Institute of Clerks of Works (Incorporated) and published by the Institute at 41 The Mall, Ealing, London W5 3TJ

PRICE: 20P PER COPY OR £5 PER 100 POST FREE

Chapter 5

Instructions and Variations

If the procedure set out in *Pre-Contract Practice* is followed it should be possible for complete sets of drawings, together with specification notes and nominated sub-contractors' and suppliers' estimates, to be available to the quantity surveyor, who will then be in a position to prepare fully described and accurate bills of quantities which can be annotated at a later stage. This in turn will mean that immediately the contract is placed the contractor can be handed all the necessary information to build the project in hand.

Architect's instructions

Despite this complete planning and documentation, however, it will probably be necessary from time to time for the architect to issue further drawings, details and instructions which are collectively known as 'Architect's Instructions'. The conditions of contract lay down those matters in connection with which the architect is empowered to issue such instructions, and as the contractor is entitled under clause 4.2 of JCT 80 to question the architect's right to issue any particular instruction, it is as well to be clear as to which clauses give him the authority. They are as follows:

Clause

2.3	Discrepancies in documents
6.1.3	Compliance with statutory requirements
7	Levels and setting out the works
8.3	Opening up work for inspection
8.4	Architects' powers when work, materials or goods not in accordance with the contract
8.5	Exclusion from the works of any person employed thereon

13.2	Variations as defined in clause 13.1
13.3.1	Expenditure of provisional sums included in the contract bills
13.3.2	Expenditure of provisional sums included in a sub-contract
17.2 ⎱ **17.3** ⎰	Making good defects, shrinkage and other faults
21.2.2	Insurance against damage caused by the carrying out of the works to property other than the works
22D.1	Insurance for employer's loss of liquidated and ascertained damages
23.2	Postponement of the execution of any works
32.2	Protection of the work in the event of war
33.1.2	Action in the event of war damage
34.2	Action in the event of antiquities being found
35.5.2	Application of alternative method of nominating a sub-contractor in lieu of the basic method
35.8	Action in the event of the contractor being unable to reach agreement with proposed nominated sub-contractor
35.9	Action in the event of proposed nominated sub-contractor withdrawing his tender
35.10.2	Nomination of sub-contractor under the basic method
35.11.2	Nomination of sub-contractor under the alternative method
35.18.1.1	Nomination of a substituted sub-contractor in the event of a nominated sub-contractor failing to rectify defects
35.23	Omission of work for which it was proposed to nominate a sub-contractor or nomination of an alternative sub-contractor if the proposed nomination does not proceed
35.24.6	Procedure following contractor's application to determine a nominated sub-contractor's employment as a result of the sub-contractor's default; and subsequent re-nomination
35.24.7	Re-nomination in the event of a nominated sub-contractor's bankruptcy etc. or if the contractor is required by the employer to determine the employment of a nominated sub-contractor
35.24.8	Re-nomination in the event that a nominated sub-contractor determines employment under the sub-contract or where, through no fault on the part of the

sub-contractor, works already carried out under the sub-contractor require to be redone, and the nominated sub-contractor refuses so to do

36.1 ⎫
36.2 ⎭ Nomination of nominated suppliers

The procedure for the issue of instructions is set out in clause 2(3) and may be summarised as follows:

(1) An instruction issued by the architect must be in writing
(2) An oral instruction is not effective unless it is confirmed by the contractor or architect within seven days. If confirmed by the contractor within the stated period and the architect does not dissent then it is deemed to be an architect's instruction, and
(3) The instruction is effective from the date of the issue of the architect's instruction or the expiration of the seven-day period referred to above, but
(4) If neither the architect nor the contractor confirms any oral instructions, but the contractor still carries out the work in question as a matter of goodwill, then the architect can at any time up to the issue of the final certificate confirm the oral instructions in writing; these instructions should then be effective from the date of the written confirmation.

It is important to note that if the clerk of works issues any directions to the contractor or to his foreman, such directions are only effective if they are in respect of matters about which the architect is empowered to issue instructions, and if they are confirmed by the architect within two working days, not, it will be noted, within seven days as provided for confirmation of the architect's own verbal instructions.

All instructions from the architect to the contractor should be issued, or confirmed, on a standard form. Such a form, shown in Example 7, is published by RIBA Publications Ltd, but many other forms, containing the same essential features, have been produced for specific jobs or architectural offices. The use of such forms has several advantages, not least being the saving in work in issuing such instructions and the fact that they are immediately recognisable to all parties to the contract.

It contributes to the smooth running of the contract if verbal instructions given by the architect during site visits or directions given by the clerk of works are recorded in a duplicating site

'an instruction issued by the architect must be in writing'

instruction book which can be signed at the time and subsequently confirmed by an instruction as set out above.

It is essential that instructions should be clear and precise, and where revised drawings are issued, the revision should be specifically referred to. Instructions emanating from consultants should be passed to the architect for confirmation by architect's instructions. Copies of architect's instructions should be distributed as follows:

- general contractor (2 copies)
- clerk of works
- quantity surveyor
- consultant (if concerned)

It should be stressed at the initial site meeting to all concerned that no adjustment will be made to the contract sum unless the matter is covered by an architect's instruction in accordance with the terms of contract.

Variations

Before considering those instructions which involve variations it might be useful to understand the meaning of 'variation' as it has been defined in tabulated form in clause 13 and as set out below.

13.1 The term 'variation' as used in the conditions means:

13.1 **.1** the alteration or modification of the design, quality or quantity of the works as shown upon the contract drawings and described by or referred to in the contract bills, including:

    .1 **.1** the addition, omission or substitution of any work;

    .1 **.2** the alteration of the kind or standard of any of the materials or goods to be used in the works;

    .1 **.3** the removal from the site of any work executed or materials or goods brought thereon by the contractor for the purpose of the works other than work materials or goods which are not in accordance with this contract;

13.1 **.2** the addition, alteration or omission of any obligations or restrictions imposed by the employer in the contract bills in regard to:

    .2 **.1** access to the site or use of any specific parts of the site;

    .2 **.2** limitations of working space;

    .2 **.3** limitations of working hours;

    .2 **.4** the execution or completion of the work in any specific order; but excludes:

13.1 **.3** nomination of a sub-contractor to supply and fix materials or goods or to execute work of which the measured quantities have been set out and priced by the contractor in the contract bills for supply and fixing or execution by the contractor.

The matters set out in clause 13.1.2 relate to some of the obligations and restrictions imposed by the employer, particulars of which will have been set out in the bills of quantities in accordance with Section A of the Standard Method of Measurement SMM7. It should be noted, however, that, while clause 4.1.1 requires the contractor to comply forthwith with all architect's instructions, those requiring a variation within the meaning of clause 13.1.2 are excepted from that general obligation.

Clearly an architect's instruction requiring a variation in

respect of the matters referred to in clause 13.1.2 could materially affect the operation of the whole contract as far as the contractor is concerned. If the contractor considers this to be the case he would not be obliged to comply with the instruction immediately. Should this situation arise the contractor would have to set out in writing his reasons for objecting to the instruction and there would have to be a separate agreement as to how the changes in the obligations and the restrictions imposed by the employer are to be dealt with. Any adjustment to the contract sum would then be made in accordance with that agreement and not dealt with as a variation under contract. If the parties are unable to reach agreement on the matter it could then be referred to arbitration, clause 41.3.3 making provision for such a dispute to be arbitrated upon immediately.

Architect's instructions nominating sub-contractors and suppliers are covered by clauses 35.10.2 and 36.2 respectively and are dealt with in more detail later in this section. It should be noted that clause 13.1.3 precludes the architect from issuing an instruction nominating a sub-contractor to carry out work which has been measured in the bills of quantities and priced by the contractor as general contractor's work. If the architect wishes to do this he may do so only with the contractor's agreement. If the contractor does agree, arrangements for his profit, attendance on the sub-contractor and any other financial matters should be settled before the nomination is made.

Clause 13.3 requires the architect to issue instructions in regard to the expenditure of provisional sums in either the main contract bills or in any sub-contract.

In summary, the architect's instruction will describe the varied work, state any item or items to be omitted, and will record any new or amended drawings which are to be worked to. If the instruction involves the adjustment of prime cost sums, particulars must be given of the estimate which is to be accepted, stating the date of the quotation and the reference number as well as the total value of the proposed accepted estimate.

Valuing variations

In the following it has been assumed that Amendment 7 to JCT 80 will have been incorporated in the contract. This amendment, issued in July 1988, relates to the preparation of the bills of quantities in accordance with the standard method of measurement SMM7.

The contract provides that variations and work carried out by the contractor in expenditure of provisional sums and work for which an approximate quantity is included in the contract bills shall be valued by the quantity surveyor and clause 13.5 comprises the rules to be observed in making the valuation, which may be summarised as follows:

- Work which can be properly valued by measurement:
    (a) Work similar to that set out in the contract bills, executed under similar conditions and with no significant change in the total quantity shall be valued at bill rates. This includes work where the approximate quantity in the contract bills is a reasonably accurate forecast of the quantity of work required.
    (b) Work of a similar nature to that set out in the contract bills but not executed under similar conditions or where there is a significant change in quantity shall be valued on the basis of those in the contract bills but with a fair allowance being made for the differences in conditions and/or quantity. This includes work where the approximate quantity in the contract bills is not a reasonably accurate forecast of the quantity of work required.
    (c) Work not similar to that set out in the contract bills shall be valued at fair rates and prices.
- Omissions shall be valued at the rates and prices contained in the bills.
- When valuing the foregoing the following must be taken into account:
    (a) Measurement must be in accordance with the method of measurement used for the preparation of the bills of quantities.
    (b) Allowance must be made for any percentage or lump sum adjustment in the contract bills.
    (c) If appropriate, allowance must be made for any addition to or reduction of preliminaries.
- Work which cannot properly be valued by measurement shall be valued on a daywork basis.
- If any variation substantially changes the conditions under which other work is executed then that other work shall be revalued as if it was itself a variation.
- If the variation does not involve additional or substituted work or straightforward omissions, or if the valuation cannot reasonably be effected by the application of these valuation rules, then a fair valuation must be made.

'. . . work of a similar nature . . .'

It is now quite clear that if a variation causes either a change in the working conditions or a significant change in the quantities or in the conditions under which other work is carried out, then such changes must be taken into account in the valuation.

It is also quite clear that allowance must be made in valuing variations for percentage or lump sum adjustments which have been made in the contract bills and, where appropriate, an adjustment must be made in respect of preliminaries.

The one matter which need not be taken into account when valuing variations is any effect which the variation may have had on the regular progress of the work, or any direct loss and/or expense which the contractor may have incurred as a result of the variation but which he is unable to recover through the valuation or any other provision of the contract. These matters are dealt with under clause 26.

The value of variations on sub–contract works, including the valuation of work carried out against provisional sums included in the sub-contract, is to be made in accordance with the relevant provisions of the sub-contract. The valuation rules in the sub-contract conditions are similar in principle to those of the main contract referred to above and it is not proposed to deal with them further here. In this connection, however, JCT 80 does clarify the position when the main contractor tenders for sub-contract works and his tender is accepted. Any variations in the sub-contract works are then valued in accordance with the terms of the sub-contract, not in accordance with the terms of the main contract.

Dayworks

In the case stated above, the daywork sheets must be submitted to the architect not later than one week following that in which the work was carried out. The architect or his authorised representative, usually the clerk of works, can within a reasonable period of time certify the accuracy of the hours recorded and the materials used and, if satisfied, can sign the sheets as a true record of the time and materials used. The daywork sheets will then be passed to the quantity surveyor who will satisfy himself that the labour, materials and plant recorded on them are reasonable for the work involved. He will also, of course, check the prices and arithmetic.

It is advisable that all daywork sheets be serially numbered, state the date when the work was carried out, specify the daily time spent on the work set against each operative's name and list the materials and plant used. The architect should check that he has issued the necessary instruction covering the work (a signed daywork sheet does not constitute an instruction) and reference to this instruction should be made on the daywork sheets concerned.

Nomination of sub-contractors

A nominated sub-contractor is defined as a sub-contractor whose selection is reserved to the architect. The architect may make his nomination in one of two ways, namely by what has become known as the basic method using full documentation, or by the alternative method which is a shorter procedure intended for use in the case of some of the minor items of work for which the architect may wish to nominate a sub-contractor.

The procedures required to effect a nomination, together with the paperwork which accompanies them, may seem on the face of it to be cumbersome and long-winded. This is not in fact the case. The provisions have been drawn up with a view to eliminating the many problems which have arisen in the past as a result of nominations being made on the basis of inadequate information and with little regard for the impact of the nomination on the programme and working arrangements of the general contractor. Professional advisers should welcome the comprehensive documentation which has been prepared in conjunction with the 1980 standard form.

Because of the importance of ensuring that the nominated sub-contractor's work can properly be integrated into the work of the main contractor, it is essential that decisions are taken in good time before tenders are invited. The preparatory work leading to a nomination ought, therefore, to be dealt with in the pre-contract stage and the detailed procedures under both the basic and the alternative method are described in *Pre-Contract Practice* but it may be helpful to summarise here the sequence of events leading up to the nomination.

The basic method involves the use of the Form of Tender and Agreement (NSC/1), the Employer/Nominated Sub-Contractor Agreement (NSC/2), the Form of Nomination (NSC/3) and the Standard Form of Sub-Contract (NSC/4). The procedure is as follows:

- The architect prepares Tender NSC/1, inserting in schedule 1 the particulars of the main contract.
- At the same time the architect prepares the Agreement NSC/2.
- The architect sends the original and two copies of NSC/1 and the original NSC/2 to the proposed sub-contractor.
- The proposed sub-contractor completes Tender NSC/1 and the two copies and signs on page 1.
- The proposed sub-contractor executes Agreement NSC/2 under hand or under seal as instructed by the architect.
- The proposed sub-contractor returns to the architect the original documents and the copies of NSC/1.
- The architect signs the Tender NSC/1 and the two copies on page 1 as 'approved' on behalf of the employer.
- The employer executes Agreement NSC/2 again either under

hand or under seal, retaining the original and sending the architect a certified true copy.

- The architect sends to the proposed sub-contractor the certified copy of the agreement.
- The architect sends the main contractor a preliminary notice of nomination, accompanied by the original Tender NSC/1 and its two copies as then completed, together with a copy of the Agreement NSC/2 for the main contractor's information.
- The main contractor checks that the details in schedule 1 of Tender NSC/1 are correct.
- The main contractor completes schedule 2 insofar as it has not already been completed by the proposed sub-contractor, deleting those items which are no longer relevant and agreeing all the remaining items with the sub-contractor.
- The contractor and sub-contractor sign schedule 2 and the contractor signs the tender itself to indicate his acceptance of it, subject only to the architect issuing the nomination instruction.
- The contractor then returns the original, and now completed, Tender NSC/1 and the copies to the architect.
- The architect issues his instructions to the contractor nominating the proposed sub-contractor using Nomination NSC/3, sending with it the original Tender NSC/1.
- At the same time the architect sends to the proposed sub-contractor a copy of Nomination NSC/3, together with a certified copy of the completed Tender NSC/1.
- The contractor and the nominated sub-contractor enter into a Sub-Contract NSC/4. Whether this will be under hand or under seal will have been stated previously in schedule 2 of Tender NSC/1.

The alternative method for nominating is used where the sub-contract works are not complicated enough for the full formalities of the basic method to be appropriate. There are two options under the alternative method: the 'without documents' option and the 'with documents' option.

The 'without documents' option is used where the sub-contract works are of a simple nature and for this the nominating process has no standard documentation. The form of sub-contract for this option is the version of NSC4a which includes the conditions of sub-contract within its 51 pages and which was issued by the JCT in 1980.

The 'with documents' option is used where the sub-contract

works are too complex for the 'without documents' option to be used but where the employer or the design team do not wish to follow the basic method. For this the tender and nomination documents NSC1a and NSC3a should be used. The form of sub-contract for this option is the five page version of NSC4a, issued by the JCT in 1986, in which the conditions of sub-contract are incorported by reference only.

The form of employer/sub-contractor agreement, NSC/2a, can be used with either option.

If it is proposed to use the alternative method this must have been stated in the bills of quantities or in the architect's instruction under which the nomination was made. At the same time it must be stated whether or not the agreement NSC/2a is to be entered into. The architect may, however, issue an instruction substituting the alternative method for the basic method, or the basic method for the alternative method, but any such instruction must be issued before the preliminary notice of nomination. Such an instruction would be treated as a variation.

Under the basic method, if the contractor and the proposed sub-contractor are unable to agree on the particular conditions to be inserted in Schedule 2 of Tender NSC/1 the contractor must inform the architect in writing, giving reasons for the inability to reach agreement. The architect must then issue such instructions as may be necessary. If the proposed sub-contractor validly withdraws his offer, again the contractor must inform the architect in writing and await the architect's instructions. Similarly under the alternative method, if the proposed sub-contractor without good cause fails within a reasonable time to enter into a sub-contract NSC/4a then the architect must issue appropriate instructions. In such circumstances the architect must either make a fresh nomination or issue an instruction requiring as a variation the omission of the work concerned. In which case he may not, without prior agreement, instruct the contractor to carry out the work.

The contractor does, of course, still have the right to make a reasonable objection to a nominated sub-contractor and if he wishes to make such an objection the contract requires him to do so at the earliest practicable moment and in any case not later than when he returns Tender NSC/1 to the architect. If the alternative method of nomination applies, under which NSC/1 is not used, then the contractor must make his objection within seven days of receiving the architect's instruction nominating the sub-contractor. It must always be borne in mind that a sub-contractor may only be nominated for work covered by a p.c. sum in the bills of quantities, where the sub-contractor is named in the bills, in an instruction regarding the expenditure of a provisional sum, or, subject to certain qualifications, in a variation order.

If, after nomination, a sub-contractor defaults in his perform-ance of the contract, or if he goes into liquidation, the architect must make a fresh nomination. Clause 35.24 sets out the procedure to be followed in such circumstances. As this is not a very common occurrence it is not proposed to enlarge on those procedures here, but they should of course be carefully followed when such a situation arises.

Nomination of suppliers

A supplier is nominated or deemed to be nominated if the supply of materials or goods is covered by a p.c. sum in the bills and the supplier is either named in the bills or subsequently named by the architect in his instruction regarding the expenditure of the p.c. sum; or where in an instruction regarding the expenditure of a provisional sum or in a variation order the architect specifies materials or goods which can only be purchased from one supplier. In the latter case the materials or goods concerned must be made the subject of a p.c. sum in the architect's instruction.

Clause 36 requires the architect to issue instructions for the purpose of nominating a supplier for any materials or goods covered by a p.c. sum and the clause sets out the manner in which the costs to be set against the p.c. sum are to be ascertained.

Clause 36.4 sets out the conditions of sale which the supplier will be required to accept in his contract of sale with the contractor and if the supplier refuses to accept those conditions the contractor cannot be required to accept the nomination. The JCT Form of Tender for nominated suppliers is Tender TNS/1 and this substantially reproduces in Schedule 2 the provisions of clauses 36.3–8.

Cost control

Cost control may be defined as the controlling measures necessary to ensure that the authorised cost of the project is not exceeded. It is a continuing process following the cost planning and cost control exercised during the design period and is discussed in *Pre-Contract Practice.*

Normally the authorised cost will be represented by the contract sum. However, there may be occasions when the sum can be varied while construction is taking place. For example, a client building speculatively may find that his tenant is prepared to pay more for a particular facility, in which case he will give instructions for this to be assessed, and, if implemented, an adjustment will be made to the authorised cost.

One essential point to establish at the outset is the extent to which the client requires cost control. Both architect and quantity surveyor should obtain the client's instructions on this point. On the one hand he may issue an overall instruction that on no account must the authorised sum be exceeded; alternatively, he may require the cost to be carefully monitored throughout while giving overriding priority to the quality of the building. In this connection it is good practice to clarify the purpose of the contingency sum with the client at the outset of the job.

Generally the contingency sum should only be used to cover the cost of extra work which could not reasonably have been foreseen at the design stage (e.g. extra work below ground level). It should not be used for design alterations, except with the prior approval of the client. The contingency sum can, however, be used to cover the cost of any insurances under 21.2 (adjoining property) and/or 22D (loss of liquidated and ascertained damages) if these have been instructed by the architect. This avoids the need to increase the contract sum if such insurances are required and also avoids the need to include provisional sums for such insurances in the contract documents. Too often in the past the latter practice resulted in insurance premiums which exactly equalled the provisional sum allowances.

To maintain cost control, it is necessary for the value of all possible variations to be assessed before instructions are issued so that their effect may be taken into account. This process needs extremely close liaison between architect and quantity surveyor and the other consultants. It demands the quantity surveyor's attendance at all relevant meetings, including site meetings, and the submission to him of copies of all correspondence which may have cost implications. Where time allows, it is a useful discipline for all proposed instructions to be discussed with the quantity surveyor for pricing, prior to their formal issue. This allows consideration to be given to the cost effect of an instruction before the expenditure is committed. It is also wise for the architect to look ahead and make early decisions on such matters as expenditure against provisional and prime cost sums and the likelihood of variations in the later stages.

The necessity for strict cost control does not eliminate the need for continuing cost studies in areas of construction which have not been finally detailed. If this is done effectively the likelihood of having to draw on the contingency sum is lessened. Indeed the contingency provision can even be built up to cope with unforeseen variations and if none occurs a saving will ultimately be made for the client against the authorised cost.

The evaluation of possible variations and the likely outcome of expenditure to be set against the prime cost and provisional sums should not preclude looking at legitimate claims which might be made by the contractor and taking these into account early on in forecasting final cost.

Whether full cost control is required or monitoring of costs, it will be necessary for the quantity surveyor to produce monthly forecasts of final expenditure. In these he should not only take into account formal architect's instructions, but also known extras or savings. For example, if the district surveyor or fire officer has indicated that he will be making certain requirements that have not been taken into account in the contract, then the quantity surveyor should include an assessment of the effects of these pending the formal instructions from the architect.

The monthly forecasts of final expenditure should take into account the amount of the contingency sum still remaining and consideration should be given to the adequacy of any such amount in the circumstances. In the early stages of the work most of the contingency sum, if it remains unspent, should still be included, but as the contract progresses it can be reduced as the risk of unforeseen extras diminishes. In the event of the contingency sum being spent in the early stages, consideration will have to be given to including a further contingency allowance in cost reports and notifying the client accordingly.

On large projects, it may well be inappropriate for monthly statements to be made and quarterly will suffice.

Where specialist engineering works are designed by a consulting engineer, instructions, although emanating from the engineer, will be issued by the architect. Cost control, preferably before the issue of these instructions, is just as important as with any other instructions and close liaison is essential so that all parties know the full implications of the costs identified.

An example of a financial report is given in Example 8.

EXAMPLE 7

Issued by: Reed & Seymore
address: 12 The Broadway
Borchester BC4 2NW

Architect's instruction

Employer: Aqua Products Ltd
address: Waterside Road, Borchester BC1 6AD

Serial no: 12

Job reference: 456

Contractor: Leavesden Barnes & Co Ltd
address: Thatchers Yard
Midley BC14 6SE

Issue date: 11 September 1989

Contract dated: 1 June 1989

Works: Shops & Offices
Situated at: Newbridge Street, Borchester

Under the terms of the above Contract, I/We issue the following instructions:

	Office use: Approx costs	
	£ omit	£ add

1. External Door Thresholds (B/Q ref 4/28a)
   To external doors DE2 DE3 DE4 & DE7:

   Omit: Galvanised water bars bedded in concrete thresholds
   Add: Extruded aluminium draught excluder thresholds;
         'Alumat' ref no 1724 dowelled & screwed to grano
         paving.

2. Pointing to Brickwork below dpc (B/Q ref 3/6d)
   Omit: Pointing struck flush as brickwork rises for facing
         brickwork below dpc only
   Add: Rake out joints ready to carry out pointing at time
         of raising brickwork above dpc.

   All pointing to be weather struck, carried out as brick
   rises above dpc.

3. Remedial Work-Ground Floor Slab
   (Confirmation of Site Instruction 13)
   Make good approx 60 m2 of improperly power floated concrete finish between grid lines A4 & C5; surface to be
   mechanically scrabbled and resurfaced with a self levelling epoxy screed laid strictly in accordance with the
   manufacturer's instructions. Copies of the specification
   of the proposed material and laying instructions to be
   sent to the structural engineer for agreement.

4. Issue of Revised Drawings
   Work to be carried out in accordance with the following
   drawings issued under cover of Drawing Issue Record
   dated 10 October 1989:
   456/W/7 rev B, 456/W/8 rev D, 456/W/16 rev A.

To be signed by or for
the issuer named
above.   Signed _____

Amount of Contract Sum	£	
± Approximate value of previous instructions	£	
	£	
± Approximate value of this instruction	£	
Approximate adjusted total	£	

Distribution [1] Employer   [1] Contractor   [1] Quantity Surveyor   [1] Services Engineer

[ ]   [ ] Nominated   [1] Structural Engineer [1] File
                      Sub-Contractors

© 1985 RIBA Publications Ltd

EXAMPLE 8: FINANCIAL REPORT

FUSSEDON KNOWLES & PARTNERS
Chartered Quantity Surveyors

Upper Market Street,
Borchester BC2 1HH

FINANCIAL REPORT NO. 9

4th March 1990

PROJECT: Shops and Offices,
            Newbridge Street, Borchester

EMPLOYER: Aqua Products Ltd.,
            Waterside Road, Borchester BC1 6AD

ARCHITECT: Reed & Seymore,
            12 The Broadway, Borchester BC4 2NW

CONTRACTOR: Leavesden Barnes & Co. Ltd.,
            Thatchers Yard, Midley BC14 6SE

CONTRACT DATES:

Possession: 10th June 1989
Completion: 11th July 1990   Extended to: 25th July 1990

CERTIFIED TO DATE: £511,186.25

			£
CONTRACT SUM:			675,332
CONTINGENCIES:			20,000
	Contract Sum excluding contingencies		655,332
VARIATIONS:	Net additions as Appendix 1	£22,005	
	Allowance for future anticipated variations	2,000	24,005
			679,337
FLUCTUATIONS:	Not applicable		-
			679,337
CLAIMS:	In respect of postponement of date of possession		2,000
	ESTIMATE OF FINAL COST		£681,337

EXAMPLE 8 *cont.*

FINANCIAL REPORT NO. 9   Appendix 1

PROJECT: Shops and Offices,
              Newbridge Street, Borchester

Instruction/ Variation	Brief Description	Estimate Omission £	Addition £
	FINANCIAL REPORT NO. 8 Brought Forward	5,000	29,961
	Adjustment to previous Instructions	NIL	NIL
	VARIATIONS this month:-		
AI 44	Additional steel beam in lift motor room.		180
AI 45	Variations to sanitary fittings in office toilets.	340	
AI 46 (1)	Omission of emulsion paint in stores.	232	
(2)	Hardwood strip flooring in area G4.		660
(3)	Revisions to suspended ceiling specification.	2,200	
AI 47	Additional ventilation ducting at east end.		310
	Adjustment for provisional quantities for work in connection with incoming service mains.	734	
(AI not yet issued)	Omission of plant tubs in forecourt.	600	
		9,106	31,111
			9,106
	Net Addition		£22,005

Chapter 6

Interim Certificates

The planning and maintaining of a satisfactory cash flow is vital to any business enterprise and this is especially true of building in which many different participants are involved, in which substantial expenditure must be incurred in advance of payment and in which large sums of money must pass regularly from one party to another if the whole process is to be viable. It is not surprising, therefore, that clause 30 of the conditions of contract which sets out the procedure for payment to the contractor deals with the subject in considerable detail.

When considering this subject it is perhaps helpful to bear in mind that an employer's contractual obligation to make regular payments to a contractor is accompanied by similar obligations of the contractor to make regular payments to his suppliers and subcontractors, nominated and domestic. Although we do not deal with those matters in this book it might be noted for reference purposes that the contractor's payment obligations to his nominated sub-contractors are set out in clause 21 of the sub-contract NSC/4 and domestic sub-contractors in clause 21 of the form DOM/1. Payments to nominated suppliers are dealt with in Clause 36.4.4. and 36.4.6 as qualified by 36.4. (These clauses are repeated in Schedule 2 of TNS/1 – the JCT Standard Form of Tender by Nominated Supplier). Payments to other suppliers are usually made monthly, but will depend on the terms which contractors may negotiate with them.

Obligations

The obligations of the parties to the contract and the employer's consultants are briefly as follows:

The employer

As soon as the architect issues an interim certificate stating that a payment is due from the employer to the contractor, the employer incurs a debt which must be paid within 14 days of the date on which the architect issues the certificate. (In respect of the issue of the final certificate the balance stated as due must be paid within 28 days of the date of issue.) If the employer fails to make payment on an interim certificate within 14 days the contractor is empowered, subject to giving note as required in clause 28.1.1, to determine his employment and he can start proceedings in the courts for the recovery of the debt. The employer might however be able to stop such proceedings by giving notice of arbitration if he disagrees with the architect's certificate. As explained in chapter 8, this is one of the matters which can be dealt with by arbitration before the practical completion of the works.

The employer does not have the right to obstruct or interfere with the issue of any architect's certificate and if this were to happen in respect of an interim certificate it is again a matter for which the contractor might determine his employment under the contract. However, if applicable, he does have the right to deduct liquidated and ascertained damages from an interim certificate. The contractor would achieve nothing by determining if it were the final certificate which was obstructed or interfered with, and it is doubtful if he would achieve much by determining after or close to practical completion. Instead he would probably be better off commencing proceedings in the courts for recovery of a debt, or if he has other complaints, commencing arbitration proceedings.

The architect

The architect must issue interim certificates at the period stated in the appendix to the conditions of the contract, normally one month. After the certificate of practical completion has been issued the architect must continue to issue interim certificates as and when further amounts are ascertained as payable to the contractor, but in such cases the only proviso as to timing is that no interim certificate shall be issued within one month of the previous interim certificate. These provisions regarding the interim certificates after practical completion continue beyond the defects liability period and after the issue of the certificate of completion of making good defects, until such time as the final account has been settled and the architect is in a position to issue the final certificate. As soon as the final accounts of all the nominated sub-contractors have been ascertained, and in any case not less than 28 days before the issue of the final certificate,

'certificates'

the architect must issue an interim certificate which includes the final amounts due to all nominated sub-contractors. In respect of this certificate the proviso that a month must have passed since the issue of the previous certificate does not apply. This provision enables all payments to sub-contractors to be cleared before the final certificate, which will then include only the outstanding amounts due to the contractor.

When issuing his certificate the architect must direct the contractor as to the amounts included for nominated sub-contractors and he must notify each nominated sub-contractor of the amount included. The rules regarding retention also require that at the date of each interim certificate a statement shall be prepared specifying separately the amount of retention on the main contractor's work and on each nominated sub-contractor's work and copies of that statement must be issued by the architect to the employer, the contractor and each of the nominated sub-contractors involved.

Subject to the exception set out in the next section (The quantity surveyor), contract conditions state that to arrive at the amount due in an interim certificate the architect may, if he considers it necessary, request the quantity surveyor to prepare an interim valuation. It is in practice the normal procedure for the quantity surveyor to prepare valuations for each certificate.

The important part interim certificates play in the smooth running of a contract cannot be over−stressed. The following particular points should be borne in mind:

- If for any reason the quantity surveyor does not produce an interim valuation prior to the date for the issue of an interim certificate, the architect must still issue his certificate at the appointed time and in accordance with the requirements of the contract.
- The architect is legally responsible for the accuracy of his certificates.
- The architect must remain independent in the issuing of certificates so as to be fair to both parties.

The quantity surveyor

The quantity surveyor's obligations under the contract as far as interim certificates are concerned depend, strictly speaking, on which clauses apply for fluctuations. If clause 38 (contribution, levy and tax fluctuations) or clause 39 (labour and materials cost and tax fluctuations) applies, it is at the architect's discretion as to whether or not interim valuations are prepared by the quantity surveyor. If clause 40 (use of price adjustment formulae) applies, interim valuations must be prepared by the quantity surveyor prior to the issue of each interim certificate. In practice, however, it is normal for valuations to be prepared by the quantity surveyor before each certificate regardless of the way in which fluctuations are adjusted.

The contractor

The contractor is under no obligation under the contract to assist in any way in the preparation of interim valuations or certificates. It is the architect's responsibility to ascertain the amount due and to issue the certificate and it is the employer's responsibility to make payment to the contractor within the 14-day period. The architect may, and probably will, have the assistance of the quantity surveyor's valuation and in practice that valuation will probably have been prepared in conjunction with the contractor. It is customary for the contractor to co-operate with the quantity surveyor in the preparation of interim valuations and this is obviously a sensible and satisfactory procedure, but it is not an obligation.

What is an obligation on the contractor is that within 17 days of the date on which the architect issues his certificate the contractor must pay to his nominated sub-contractors the amounts due to them as directed by the architect in the interim certificate. The contractor should by then, of course, have received from the employer the amount due under the certificate, but even if that payment has not been received, the contractor's obligation to pay his nominated sub-contractors within the 17-day period remains.

Interim valuations

As previously stated, if clause 40 applies the quantity surveyor has a contractual obligation to carry out interim valuations; but even if this is not the case, monthly interim valuations are usual. Again, although it is not a contractual requirement, it is normal practice for the contractor to co-operate with the quantity surveyor in preparing the valuations.

The gross value to be included in interim certificates is spelt out in clause 30.2 under two headings and may be summarised as follows.

Amounts to be included which are subject to retention:

- the total value of the main contract works *properly executed* including variations and, where applicable, adjustment in respect of fluctuations where the price adjustment formula applies
- the total value of materials and goods delivered to or adjacent to the works
- the total value of materials and goods off site if authorised by the architect
- in respect of each nominated sub-contractor the total value of the sub-contract works, materials and goods as set out in the three points above for the main contract works

- the profit of the contractor upon the total amount included for each nominated sub-contractor including the sub-contractor's fluctuations and any other monies due to the sub-contractor under the terms of his sub-contract

Amounts to be included which are not subject to retention:

- any amount which may become due to the contractor under the terms of the contract in respect of statutory fees and charges, setting out the works, opening up and testing, royalties, remedial works where the architect authorises payment, insurances under clauses 21.2.3 and costs to the contractor if the employer defaults in insuring under clauses 22B.2, or 22C.1 and 22C.2.
- costs in respect of insurances taken out by the contractor under clause 22D.4
- any amount due to the contractor by way of reimbursement for loss and expense arising from matters materially affecting the regular progress of the works or from the discovery of antiquities
- any final payment to a nominated sub-contractor
- any adjustments under the fluctuations provisions, other than where the price adjustment formula applies
- any amount properly payable to a nominated sub-contractor in respect of statutory fees and charges, remedial works for which the architect authorises payment, and fluctuations other than where the price adjustment formula applies

The inclusion of unfixed materials and goods in interim valuations sometimes presents problems. It should be noted that materials on site, that is, stored on the site itself or, subject to the architect's approval, at a storage area nearby, must be included in the valuation at their full value unless they have been brought onto the site prematurely. For example, unless there was some prior agreement on the matter, the architect is entitled to withhold payment for, say, joinery fittings brought onto the site when site stripping is still in progress.

The inclusion in the gross value of an interim certificate of off-site materials and goods is entirely at the discretion of the architect and that discretion is now frequently exercised, usually in the case of items prepared or manufactured for a contract away from the site and stored at the supplier's or sub-contractor's premises until it is timely for them to be delivered to site. When the architect decides that such goods are to be included in interim valuations certain conditions must apply in order to safeguard the employer's interest in them. These conditions are set out comprehensively in clause 30.3 and it is the architect's responsibility to ensure that they have been complied with before he authorises payment. For that reason it is important that early decisions should be made regarding payments for off-site materials and goods so that the requirements of clause 30.3 can be complied with before the relevant interim certificate is due.

When materials and goods are certified and paid for, the contractor remains responsible for loss or damage to them. However, they become the employer's property and must not be removed from the works or place of storage without his authority. Materials on site must be adequately protected and the architect can refuse to include them in interim valuations if he is not satisfied with the protection provided. Similarly he will exercise his discretion by refusing payment for off-site materials if he considers the arrangements for their storage are not satisfactory.

Returning to the interim valuation itself, the gross valuation for an interim certificate comprises the total amounts which are subject to retention, plus the total amounts which are not subject to retention, less any amount allowable to the employer in respect of fluctuations, all as summarised above. From this gross valuation there will be deducted retention and the total amounts stated as due in interim certificates previously issued. It must always be borne in mind that any defective work must be excluded.

Retention

The contract provides for the employer to retain a percentage of the gross value included in interim certificates and clauses 30.4 and 30.5 set out in detail the rules regarding this retention.

Unless a lower rate has been agreed between the parties and stated in the appendix to the conditions, the amount retained will be 5 per cent. A footnote recommends that where a contract sum is £500 000 or more, the retention should not exceed 3 per cent. In principle this retention is ultimately released in two stages, half on the practical completion of the work and the balance when all defects have been made good at or after the end of the defects liability period. However, the contract provides for the early release of retention if the employer takes possession of part of the works or when the sectional completion supplement applies. In addition there are also provisions for retention to be released early to nominated sub-contractors. To apply these contractual provisions, therefore, it is necessary to divide the value of the work subject to retention included in each interim certificate into three parts:

- Total value of work which has not yet reached practical completion plus unfixed materials and goods } subject to full retention
- Total value of work which has reached practical completion but for which a certificate of completion of making good defects has not been issued } subject to half retention
- Total value of work for which a certificate of completion of making good defects has now been issued } nil retention

These provisions apply equally to work carried out by nominated sub-contractors, the value of which is, of course, included in the gross valuation. In the case of nominated sub-contractors there is in the normal course of events the additional obligation on the employer to make early final payment to the nominated sub-contractor in accordance with the provisions of clause 35.17 of JCT 80. That obligation arises under the terms of the Employer/Nominated Sub-Contractor Agreement (NSC/2 or NSC/2a), which now forms part of the standard documentation required for the nomination of the sub-contractor.

Clause 35.16 states that when in the opinion of the architect practical completion of the works executed by a nominated sub-contractor is achieved,he shall forthwith issue a certificate to that effect and practical completion of such works shall be deemed to have taken place on the day named in that certificate. Clause 35.17 states that where the Agreement NSC/2 or NSC/2a has been entered into and the relevant clauses of the agreement are unamended, then at any time after the date of practical completion of the sub-contract works the architect may, and on the expiry of 12 months from the date of practical completion of that work must, include in an interim certificate the final payment to the nominated sub-contractor, subject to the sub-contractor having remedied any defects and having sent to the architect or

quantity surveyor all documents necessary for the final adjustment of the sub-contract sum.

This means that the retention in respect of each nominated
sub-contractor must be separately identified in the valuation for
the interim certificate.

This separate identification of retention monies is also necessary to enable the employer's obligations regarding retention
monies to be fulfilled.

These obligations are set out in the rules on the treatment of
retention contained in clause 30.5 of JCT 80. These rules state:

(1)   That the employer's interest in retention is fiduciary as a
      trustee for the contractor and for any nominated sub-
      contractor, but without an obligation to invest. In other
      words it is money held by the employer on trust.
(2)   That at the date of each interim certificate the architect or the
      quantity surveyor shall prepare a statement specifying the
      amount of retention for the contractor and for each nomin-
      ated sub-contractor, copies of that statement being issued
      by the architect to the employer, the contractor and each
      nominated sub-contractor.
(3)   That the employer shall at the request of the contractor or
      any nominated sub-contractor place the retention money in
      a separate bank account and certify to the architect, with a
      copy to the contractor, that the retention has been so placed
      (this rule does not apply in the local authority edition of JCT
      80).
(4)   Where the employer exercises his right to deduct monies
      due to him under the terms of the contract from monies cer-
      tified as being due to the contractor he may make such a
      deduction from amounts payable under a certificate, includ-
      ing any retention released in the certificate, but he cannot
      take into account retentions still held. If a deduction is made
      the employer must inform the contractor of the reasons for
      it, and if the deduction is from retention he must state the
      amount he has deducted from the contractor's share of the
      retention or from any nominated sub-contractor's share.

It should be noted that releases of half retention on the main
contract should be included in the next interim certificate after
practical completion of the works and similarly half the retention
held on parts of the works should be released when such parts
are taken possession of by the employer under clause 18.It is
necessary for the Certificate of Practical Completion to be issued

before half the retention on the whole works may be released, but release of half the retention on a part of the works taken possession of by the employer does not require such a certificate. The early release of half retention to nominated sub-contractors on the other hand does require the issue of a certificate by the architect to the effect that practical completion of the nominated sub-contractor's work has taken place. As far as the release of the balance of retention is concerned, this is dependent on the issue by the architect of a certificate of making good defects in all cases except, of course, where there are no defects, in which case the balance of retention must be included in the next interim certificate after the expiration of the defects liability period.

Payments to nominated sub-contractors

When issuing his certificate the architect must direct the contractor as to the amounts included for nominated sub-contractors and he must notify each nominated sub-contractor of the amount included. The contractor must discharge his payments to the nominated sub-contractors within 17 days of the date of issue of the certificate.

The procedures for the nomination of sub-contractors under JCT 80, incorporating as they do the Employer/Nominated Sub-Contractor Agreement, impose strong disciplines to be observed by all concerned in this payment process.

Before the issue of each interim certificate the contractor must provide the architect with reasonable proof that he has made any payments to nominated sub-contractors due under previous certificates. If he is unable to provide such proof then the procedures for direct payment will apply, unless the contractor is able to satisfy the architect that the absence of proof is due to some failure or omission of the nominated sub-contractor concerned.

The procedure for direct payment is mandatory on the employer if the Employer/Nominated Sub-Contractor Agreement (NSC/2 or NSC/2a) has been entered into. Otherwise it is optional at the employer's discretion.

If the reasonable proof of payment is not forthcoming the architect must issue a certificate to that effect stating the amount in respect of which the contractor has failed to provide such proof. He must send a copy of that certificate to the nominated sub-contractor concerned. Provided the certificate has been issued, the amount of any future payment to the contractor will be reduced by the amount by which the contractor has defaulted in his payment to the nominated sub-contractor and the employer will pay the sub-contractor concerned direct.

This direct payment will be made at the same time that the employer makes his payment to the contractor under the next interim certificate. If there is no money due to the contractor under the next interim certificate, e.g. if the direct payments equal or exceed the amount of the certificate, then the direct payment to the sub-contractor must be made within the 14-day period within which the contractor would have been paid.

There is no obligation on the employer to make direct payments to nominated sub-contractors in excess of the amount due to the contractor. If the amount due to the contractor is retention which is being released, then the amount by which the payment to the contractor is reduced in respect of a direct payment must not exceed the contractor's share of the released retention.

If there is more than one nominated sub-contractor to be paid direct and the monies due to the contractor are insufficient to meet the total of the direct payments, then the employer must apportion the money available pro rata or on some other fair and reasonable basis.

If the contractor goes into liquidation these provisions for direct payment to nominated sub-contractors immediately cease to have effect.

Value added tax

Clause 15.2 makes it quite clear that the contract sum is exclusive of value added tax and no VAT should therefore be included in any certificate issued by the architect. The responsibility for the payment of this tax is that of the contractor and he in turn will invoice the employer appropriately. Clause 15.2 gives the contractor the right to recover VAT from the employer and the provisions for that recovery are set out in the supplement to the conditions of contract known as the VAT agreement.

It is recommended that interim and final certificates issued by the architect should state that VAT is excluded and neither the architect nor the quantity surveyor should put themselves in a position of certifying the amount of tax due. However, the employer may require advice as to whether the value of work on which the contractor is charging VAT is correct. Most architects and surveyors will assist the employer with this matter, but that would be by way of separate agreement and not as part of their duties under the contract.

Valuation and certificate forms

Although a few employers have their own forms on which they require interim payments to be certified, it is now the general practice for standard forms published by the professional bodies to be used.

Valuation forms are published by the RICS for the use of quantity surveyors. These comprise the valuation shown in Example 9 and the statement of retention and of nominated sub-contractors' values, which accompanies the valuation form as an appendix and is shown in Example 10.

The forms are published by the RIBA for the use of architects in connection with interim certificates, these being the interim certificate, the statement of retention and of nominated sub-contractors' values and the notification to nominated sub-contractors concerning amounts included for them in the certificate. The first two of these forms can be filled in directly from the information given in the quantity surveyor's valuation and completed specimens are shown in Examples 11 and 12. The notification form is shown in Example 13.

All these forms are published in separate pads and each pad contains notes on their application. It will also be seen that the quantity surveyor's valuation form contains several notes which form a useful reminder of the contractual obligations.

EXAMPLE 9

Valuation

Chartered Quantity Surveyor
Fussedon Knowles & Partners,
Upper Market Street,
Borchester BC2 1HH

Valuation No: 9
Date of issue: 4th March 1990
QS Reference: B/8266

To Architect/Contract Administrator

Reed & Seymore
12 The Broadway
Borchester BC4 2NW

Employer

Aqua Products Ltd.
Waterside Road
Borchester BC1 6AD

Contractor

Leavesden Barnes & Co. Ltd.
Thatchers Yard
Midley BC14 6SE

Contract sum £ 675,332.00

Works
Shops and Offices
Newbridge Street
Borchester

I/We have made, under the terms of the Contract, an Interim Valuation

as at **4th March 1990** * and I/we report as follows:—

Gross Valuation
(excluding any work or material notified to me/us by the Architect/The Contract Administrator in writing, as not being in accordance with the Contract).

£ 536,786.00

Less total amount of Retention, as attached Statement.

£ 25,599.75

£ 511,186.25

Less total amount stated as due in Interim Certificates previously issued by the Architect/The Contract Administrator up to and including Interim Certificate No. 8

£ 435,401.80

Balance (in words) **Seventy-Five Thousand, Seven Hundred and Eighty-Four Pounds and 45 pence**

£ 75,784.45

Signature: Chartered Quantity Surveyor FRICS/ARICS
(delete as applicable)

Notes:
(i) All the above amounts are exclusive of V.A.T.
(ii) The balance stated is subject to any statutory deductions which the Employer may be obliged to make under the provisions of the Finance (No. 2) Act 1975 where the Employer is classed as a 'Contractor' for the purposes of the Act.
(iii) It is assumed that the Architect/The Contract Administrator will:—
  (a) satisfy himself that there is no further work or material which is not in accordance with the Contract.
  (b) notify Nominated Sub-Contractors of payments directed for them and of Retention held therein by the Employer.
  (c) satisfy himself that the previous payments directed for Nominated Sub-Contractors have been discharged.
* (iv) The Architect's/The Contract Administrator's Interim Certificate should be issued within seven days of the date indicated thus
(v) Action by the Contractor should be taken on the basis of figures in, or attached to, the Architect's/The Contract Administrator's Interim Certificate.

© 1980 RICS

EXAMPLE 10

Statement of Retention and of Nominated Sub-Contractors' Values

Quantity Surveyor
Fussedon Knowles & Partners
Upper Market Street
Borchester BC2 1HH

Works
Shops and Offices
Newbridge Street, Borchester

This Statement relates to:
Valuation No: **9**
Date of issue: **4th March 1990**
QS Reference: **B/8266**

	Gross Valuation £	Amount subject to: Full Retention of 5 % £	Half Retention of 2½ % £	No Retention £	Amount of Retention £	Net Valuation £	Amount Previously Certified £	Balance £
Main Contractor	256,181.84	256,031.84	–	150.00	12,801.59	243,380.25	202,758.25	40,622.00
Nominated Sub-Contractors:-								
Borchester Steelworks Ltd.	49,282.00	–	49,282.00	–	1,232.05	48,049.95	47,823.75	226.20
Midlands Claddings Ltd.	74,800.00	74,800.00	–	–	3,740.00	71,060.00	70,015.00	1,045.00
J. V. Hickson Ltd.	55,500.00	55,500.00	–	–	2,775.00	52,725.00	43,062.55	9,662.45
Watts Fabrications Ltd.	4,463.00	4,463.00	–	–	223.15	4,239.85	4,239.85	Nil
Torch Welding Ltd.	3,650.00	3,650.00	–	–	182.50	3,467.50	3,325.00	142.50
H. & C. Piper Ltd.	58,000.00	58,000.00	–	–	2,900.00	55,100.00	39,900.00	15,200.00
Smith & Jones (Electricals) Ltd.	26,325.00	26,325.00	–	–	1,316.25	25,008.75	16,458.75	8,550.00
Trundle Doors Ltd.	7,954.00	7,954.00	–	–	397.70	7,556.30	7,220.00	336.30
Flashman Lightning Engineers	630.16	630.16	–	–	31.51	598.65	598.65	Nil
TOTAL	536,786.00	487,354.00	49,282.00	150.00	25,599.75	511,186.25	435,401.80	75,784.45

No account has been taken of any discounts for cash to which the Contractor may be entitled if discharging the balance within 17 days of the issue of the Architect/S.O.'s Certificate. The sums stated are exclusive of V.A.T.

© 1980 RICS

EXAMPLE 11

Issued by: Reed & Seymore
address: 12 the Broadway
Borchester BC4 2NW

Interim certificate
and Direction

Employer: Aqua Products Ltd
address: Waterside Road
Borchester BC1 6AD

Serial no: **B** 616263

Interim Certificate no: 9

Contractor: Leavesden Barnes & Co Ltd
address: Thatchers Yard
Midley BC14 6SE

Job reference: 456

Issue date: 6 March 1990

Works: Shops & Offices
Situated at: Newbridge Street, Borchester

Valuation date: 4 March 1990

Original to Employer

Contract dated: 1 June 1989

Under the terms of the above mentioned Contract, in the sum of

£ 675,332.00

I/We certify that interim payment as shown is due from the Employer to the Contractor, and

I/We direct the Contractor that the amounts of interim or final payments due to Nominated Sub-Contractors included in this Certificate and listed on the attached *Statement of Retention and of Nominated Sub-Contractors' Values* are to be discharged to those named.

Gross valuation inclusive of the value of Works by Nominated Sub-Contractors	£ 536,786.00
Less Retention which may be retained by the Employer as detailed on the Statement of Retention	£ 25,599.75
Sub-total	£ 511,186.25
Less total amount stated as due in Interim Certificates previously issued up to and including Interim Certificate no: 8	£ 435,401.80
Amount for payment on this Certificate	£ 75,784.45

(in words) Seventy five thousand, seven hundred and eighty four pounds - 45p

All amounts are exclusive of VAT

To be signed by or for the issuer named above.

Signed _____

Contractor's provisional assessment of total amounts included in above certificate on which VAT will be chargeable £_____ @____ %

This is not a Tax Invoice

© 1985 RIBA Publications Ltd

EXAMPLE 12

Statement of retention and of Nominated Sub-Contractors' values

Issued by: Reed & Seymore
address: 12 The Broadway, Borchester BC4 2NW

Relating to Certificate no: 9
Job reference: 456

Works: Shops & Offices
Situated at: Newbridge Street, Borchester

Issue date: 6 March 1990

	Gross valuation £	Amount subject to:			Amount of retention £	Net valuation £	Previously certified £	Balance due £
		Full retention of 5 % £	Half retention of 2½ % £	Nil retention £				
Main Contractor:	256,181.84	256,031.84	–	150.00	12,801.59	243,380.25	202,758.25	40,622.00
Nominated Sub-Contractors:								
Borchester Steelworks Ltd	49,282.00	–	49,282.00	–	1,232.05	48,049.95	47,823.75	226.20
Midland Claddings Ltd	74,800.00	74,800.00	–	–	3,740.00	71,060.00	70,015.00	1,045.00
J V Hickson Ltd	55,500.00	55,500.00	–	–	2,775.00	52,725.00	43,062.55	9,662.45
Watts Fabrications Ltd	4,463.00	4,463.00	–	–	223.15	4,239.85	4,239.85	NIL
Torch Welding Ltd	3,650.00	3,650.00	–	–	182.50	3,467.50	3,325.00	142.50
H & C Piper Ltd	58,000.00	58,000.00	–	–	2,900.00	55,100.00	39,900.00	15,200.00
Smith & Jones (Electricals) Ltd	26,325.00	26,325.00	–	–	1,316.25	25,008.75	16,458.75	8,550.00
Trundle Doors Ltd	7,954.00	7,954.00	–	–	397.70	7,556.30	7,220.00	336.30
Flashman Lightning Engineers	630.16	630.16	–	–	31.51	598.65	598.65	NIL
Total (The sums stated are exclusive of VAT)	536,786.00	487,354.00	49,282.00	150.00	25,599.75	511,186.25	435,401.80	75,784.45

No account has been taken of any discounts for cash to which the Contractor may be entitled if discharging the balance within 17 days of the issue of the Architect/Supervising Officer's Certificate.

EXAMPLE 13

Issued by: Reed & Seymore
address: 12 The Broadway
Borchester BC4 2NW

Employer: Aqua Products Ltd
address: Waterside Road, Borchester BC1 6AD

Serial no: 9

Contractor: Leavesden Barnes & Co Ltd
address: Thatchers Yard
Midley BC14 6SE

Job reference: 456

Issue date: 6 March 1990

Works: Shops & Offices
Situated at: Newbridge Street, Borchester

Valuation date: 4 March 1990

Contract dated: 1 June 1989

Nominated Borchester Steelworks Ltd
Sub-Contractor: Cannery Lane, Borchester BC7 3SW
address:

Original to Nominated
Sub-Contractor

*Delete as appropriate

Under the terms of the above mentioned Contract, I/we inform you that an interim/~~a final~~* payment due to you of £ 226.20 has been included in Interim Certificate no. 9 dated 6.3.90

issued to the Employer, and that the Contractor named above has been directed in the said Certificate to discharge his obligation to pay this amount in accordance with the terms of the Contract and the relevant Sub-Contract. Where applicable, retention has been deducted but no account has been taken of any discounts for cash to which the Contractor may be entitled.

To comply with your obligation to provide written proof of discharge of the certified amount, you should return the acknowledgement slip below to the Contractor immediately upon such discharge.

To be signed by or for the issuer named above.

Signed

Contractor:
address:

Job reference:

Works:
Situated at:

Notification date:

Interim Certificate Serial no:

We confirm that we have received from you discharge of the amount

included in Interim Certificate Serial no: _____ dated _____

as stated in the Notification dated _____

in accordance with the terms of the relevant Sub-Contract.

Signed _____ Nominated Sub-Contractor

Date _____

© 1985 RIBA Publications Ltd

Chapter 7

Completion, Defects and the Final Account

Whilst a building contract does not finally come to an end until the architect issues his final certificate, completion of the works may take place in three stages:

- Practical completion
- Sectional completion, if the sectional completion supplement (Practice Note 1) applies
- Completion of making good defects

Practical completion

Practical completion of the works is defined in clause 17 of the conditions of contract. It is the stage of the contract at which, in the opinion of the architect, the works including authorised variations are complete and have been carried out in accordance with the contract documents. This date has an important influence on the contract and any legal dispute which may arise.

Very often the architect finds that the building is ready for occupation although a number of minor items have not been completed by the contractor. The implications for the employer and the contractor in issuing the certificate of practical completion at this point must be considered by the architect – for example, the contractor's difficulty in returning to finish incomplete works in a building occupied by the employer, purchasers or tenants. If the architect is prepared to issue his certificate of practical completion he would be well advised to separately identify in detail to the contractor items which require immediate action. This may also apply to defective work which the architect requires to be remedied immediately. It is important that the outstanding works are of a very minor nature because it is clear in the contract that the retention fund remaining after issue of this certificate is only intended to cover defects discovered after practical completion and not to provide for future payment of incomplete works.

As soon as he has reached his decision as to whether the works are practically complete, the architect is required to issue a certificate stating the date of practical completion. A copy of the RIBA Certificate of Practical Completion is shown in Example 14.

When practical completion has been achieved various provisions of the contract take effect. Some of these are dependent on the issue of the architect's certificate and others are not. These distinctions are important and they arise because it is rarely feasible for the architect to reach his decision and issue his certificate on the date of practical completion itself.

In that first interim certificate after practical completion, one half of the retention money held in respect of the completed works must be released to the contractor.

On the date of issue of the certificate of practical completion the contractor is relieved of his obligations to insure the works in accordance with clause 22A, assuming clauses 22B and 22C have been deleted. He is also no longer obliged to maintain the optional insurance under clause 22D against employer's loss of liquidated and ascertained damages.

On the date named in the certificate of practical completion the defects liability period commences, as does the period of six months in which the contractor must provide to either the architect or the quantity surveyor all documents necessary to compute the final account.

When practical completion has been reached, or is alleged to have been reached, arbitration proceedings on matters which cannot be pursued prior to practical completion can be opened.

Sectional completion

Sectional completion arises from a planned programme of work which is incorporated into the contract documents by use of the sectional completion supplement. This is set out in Practice Note 1 and involves a series of amendments to JCT 80. The supplement includes an appendix which replaces the appendix in JCT 80. Before the contract is placed, particulars of each section must be inserted in the appendix. These include for each section the value of work, dates for possession and completion, the defects liability period and the amount of liquidated and ascertained damages.

The architect is required to issue a certificate of practical completion at the completion of each section and the certificate for the final section will be his certificate of practical completion for the whole contract.

Effectively each section is dealt with as a contract within a contract and thus at practical completion of a section the contractor is entitled to the release of half the retention fund for that section. Similarly the defects liability period and the period for production by the contractor of documents necessary to compute the final account commence, and the contractor's obligation to insure the section of the works under clause 22A ends, assuming clause 22A insurances applied. The contractor's obligation to maintain insurance against loss of liquidated and ascertained damages (clause 22D) for the section also ceases, assuming such insurance has been taken out. Whether or not arbitration proceedings can be commenced is optional, being a matter which must be decided before the contract is executed and the decision entered in the appendix clause against 41.2.

If the contract is large it is quite feasible to have a situation arising where a certificate of practical completion is issued on one section where the other sections have not started. In addition, final certificates can be issued on sections when other sections are not finished. The certificate to be issued by the architect in these circumstances is the same for practical completion (Example 14) suitably amended.

Partial possession

The difference between sectional completion and partial possession has been explained in Chapter 2. Sectional completion arises from a planned programme of phased completion; partial possession results from an ad hoc situation covered by clause 18

If the employer, by agreement with the contractor, takes possession of a part of the building at a date before practical completion of the whole of the works, then practical completion of that part is deemed to have occurred on that date. The architect is required, immediately after possession being taken, to issue a statement giving his approximate estimate of the value of this part, and the date of possession. The contractor is then entitled to one half of the retention fund in direct ratio to the architect's valuation of the part of the building occupied in proportion to the total value of the contract. Likewise the defects liability period commences for that part of the building at the date given in the architect's statement. This statement for partial possession only applies to the release of one half of the retention fund and the commencement of the defects liability period in relation to the relevant part. It does not affect the period of final measurement nor the procedures for arbitration which rely solely on the practical completion of the whole of the works.

At the time of partial possession of any part of the building, the value of insurance and the amount of liquidated damages for non-completion must be reviewed in accordance with clause 18.1 of the contract. Here again the adjustment reduction is in direct ratio to the valuation of the building work handed over in proportion to the total value of the contract.

Possession of the building

When practical completion is reached, the architect immediately issues a certificate to that effect and the contractor ceases to be responsible for the site and relinquishes possession to the employer. At this stage the architect and contractor should arrange a handing-over meeting with the employer and ensure that he or his staff fully understand the operation of any equipment in the building such as heating plants, lifts, electric fuses and the like. In addition, certain items of the building may require a special procedure for maintenance or repair and this information should be passed to the employer in a precise manner and should be confirmed in writing. Also, as the possession of the works changes hands the responsibility for its insurance, assuming clause 22A and not 22B or 22C applied, third-party liability and insurance generally must be taken over by the employer.

A short time before practical completion and possession of the building the architect should ensure that the contractor has removed all items of plant and surplus materials. Copies of drawings showing the building as constructed, including services drawings, cross referenced to the labels and marking of pipes and the like, should be given to the employer. A further copy of these drawings could be deposited with a third party, say the bank of the employer, so that in the event of a fire a full set of proper records of the building would be available for use in its repair or reconstruction.

Defects and making good

During the defects liability period any defects, shrinkages or other faults which appear in the building and which are due to materials or workmanship not in accordance with the contract or due to frost occurring before practical completion, are to be specified by the architect in a schedule of defects.

The method adopted in preparing this schedule is usually twofold. In the first place the architect should advise the employer to keep a proper record of all defects that appear from time to time. Some definite arrangements should be agreed so that any serious defect requiring immediate attention would be reported at once to the architect while items of a minor nature could be recorded in list form. Secondly, at the completion of the defects liability period the architect in conjunction with those consultants responsible to him for the supervision of the contract should carry out detailed inspections of the building. From these inspections and from the information gained from the employer, the schedule of defects is prepared and clause 17.2 requires that it is delivered to the contractor within fourteen days after the expiry of the defects liability period.

The contractor has to rectify the items in the schedule of defects within a reasonable time and at his own expense. This equally applies to any item of defect which may occur during the defects liability period and which, in the opinion of the architect, should have immediate attention. If the contractor fails to remedy the defective items within a reasonable time the architect may instruct another contractor to do the work and deduct the charge of this second contractor from the total of the final account.

At the time these maintenance items are being dealt with the architect may be tempted to ask the contractor to carry out some minor additional works. The contract has no proper facility for this and if delays to completion have occurred, the result can be to rob the employer of his right to deduct liquidated and ascertained damages for delay by the contractor. It is also open to the contractor to raise an objection under clause 4.2 to any such purported instruction of the architect. If it is really necessary for such additional works to be carried out by the contractor, and he agrees, it must be made clear that it is outside the original contract. An exchange of letters between the employer or the architect as his agent and the contractor should suffice. Care must be taken, however, to ensure that all necessary contractual matters are dealt with in such letters.

When the architect is satisfied that all defective work has been made good he must issue a certificate to that effect. The standard form for this purpose issued by the RIBA is shown in Example 15 and is called the certificate of completion of making good defects. Before so doing the architect will be wise to ensure that the contractor has forwarded all the necessary record drawings and maintenance manuals of specialist subcontractors (the requirement for these drawings and manuals having been set out in both the main contract and relevant subcontract documents, and if necessary reiterated in the schedule of defects).

Final account

The responsibilities of the contractor and the quantity surveyor in connection with the final account are set out at the beginning of clause 30.6. These may be summarised as follows:

(1)   Not later than six months after practical completion the contractor shall send out to the architect or quantity surveyor all the documents necessary for preparing the final account.
(2)   Within three months of the contractor complying with that obligation, the architect, or if so instructed the quantity surveyor, shall ascertain any outstanding claims for loss and expense and the quantity surveyor shall prepare a statement of all other adjustments to be made to the contract sum. The ascertainment and the statement are commonly jointly referred to as the final account.
(3)   When the final account has been completed the architect must send a copy of it to the contractor and the relevant extracts therefrom to each nominated sub-contractor.

Clause 30.6 goes on to set out in tabulated form all those matters which must be dealt with in the final account in order to adjust the contract sum in accordance with the conditions. These may be summarised as follows:

To be deducted

• prime cost sums and amounts in respect of nominated sub-contractors, together with contractor's profit
• any amounts included in interim certificates and paid for by the employer in respect of any defective work by a nominated subcontractor, where the employment of the nominated

Final account

subcontractor was subsequently determined, together with contractor's profit
- provisional sums and the value of work included against approximate quantities
- variations which are omissions including, where appropriate, omissions in respect of other works carried out under changed conditions as a result of variations
- amounts allowable to the employer under the fluctuations clauses
- any deductions from the contract sum as a result of the architect not requiring contractor's errors in setting out or defects to be made good
- any other amount which is required by the contract to be deducted from the contract sum

To be added

- the total amounts of nominated sub-contracts finally adjusted in accordance with the relevant sub-contract conditions

- where the contractor has tendered for work which was to have been carried out by a nominated sub-contractor and his tender has been accepted, the amount of that tender adjusted in accordance with the terms of the tender
- the final amounts due to the nominated suppliers, including cash discounts of 5%, but excluding value added tax
- the contractor's profit on the previous three adjustments
- any amounts payable by the employer in respect of statutory fees and charges, opening up and testing, royalties, and insurances under clause 21.2.3. (and presumably 22D.4)
- additions in respect of variations including, where appropriate, additions in respect of other works carried out under changed conditions as a result of variations
- the value of work carried out against provisional sums or approximate quantities included in the contract bills
- any amounts payable by the employer to the contractor by way of reimbursement for loss and expense arising from matters materially affecting the regular progress of the works or from the discovery of antiquities
- any amount expended by the contractor as a result of loss or damage by fire or other perils where those risks are insured by the employer and the contractor is entitled to recover such amounts
- any amount payable to the contractor under the fluctuation clauses
- any other amount which is required by the contract to be added to the contract sum

The foregoing items should be shown separately and clearly in the final account and the amount of each variation and the amount due to each nominated sub-contractor and nominated supplier should be given. In preparing the final account the surveyor must provide all reasonable facilities for the contractor to be present when measurements and details are taken or recorded. In this connection it is usual and advisable for the surveyor to agree with the contractor a suitable programme and procedure for measurement so that as the document is prepared it is agreed step by step.

The contract provides for a period of final measurement and the quantity surveyor is required to complete his final account and present it to the contractor within that period. However, the draft final account is a very effective implement in maintaining cost control of the contract if it is started as the building work begins, and is, at regular intervals brought up to date as far as possible. From this draft account a report on known and anticipated expenditure can be prepared and submitted to the employer or the architect at, say, monthly intervals as a financial forecast of the probable final cost. This also enables the account to be finalised soon after the actual building work is complete and ensures that the interim payments to the contractor are realistic in relation to the value of the work carried out.

Much could be written on the detailed work required in the preparation of final accounts but in this book we are more concerned with the procedure. Safe to say, however, that delays in settlement of the final account are a cost to the contractor, and in most cases the employer also is keen to know his ultimate financial commitment. The architect, surveyor and consultants should always remember that they have a contractual duty in relation to the time for completion of the account, and the contractor should give every assistance in the prompt accounting for sub-contractors' and suppliers' accounts, agreements of measurements and prices and similar matters.

It is important that the measurements and financial assessment of variations should be completed as speedily as possible. Clause 13.2 requires that the financial adjustment of the contract sum in respect of variations should be taken into account when preparing interim certificates. Similarly loss and expense under clauses 26.1 and 34.3 and the value of works in repairs, removing debris etc. under clauses 22B.3.5 and 22C.4.4.2, should also be included in interim certificates. The provisions of these clauses should be observed as it is bad practice for the contractor to have levied against him a hidden retention for which no financial allowance is made until the final account is completed.

Final certificate

The final certificate must be issued by the architect within two months of the end of the defects liability period, or of the completion of making good defects, or of the receipt by the contractor of the completed final account, whichever is the latest date. It will usually be on the standard form shown in Example 16.

The amount of the final certificate is the amount of difference between the total of the final account and the amounts previously stated as due under interim certificates. The clause preserves the contractor's rights to any money previously certified which the employer may not have paid. The provision referred to in Chapter 6 which requires the architect to issue an interim certificate which includes the final amounts due to all nominated sub-contractors not less than 28 days before the issue of the final certificate, means that the final certificate itself would not include any further payments in respect of nominated sub-contractors. However, the architect must notify all nominated sub-contractors of the date of issue of the final certificate.

EXAMPLE 14

Issued by: Reed & Seymore
address: 12 The Broadway, Borchester BC4 2NW

Employer: Aqua Products Ltd
address: Waterside Road, Borchester BC1 6AD

Contractor: Leavesden Barnes & Co Ltd
address: Thatchers Yard, Midley BC14 6SE

Works: Shops & Offices
Situated at: Newbridge Street, Borchester

Contract dated: 1 June 1989

Certificate of

Practical Completion

of the Works

Serial no: 1

Job reference: 456

Issue date: 14 August 1990

Under the terms of the above mentioned Contract,

I/We certify that Practical Completion of the Works was achieved on:

12 August   19 90

To be signed by or for the issuer named above.

Signed

The Defects Liability Period will therefore end on:

12 February   19 91

Distribution	Original to:	Duplicate to:	Copies to:	
	[1] Employer	[1] Contractor	[1] Quantity Surveyor	[1] Services Engineer
		[1] Nominated Sub-Contractors	[1] Structural Engineer	[1] File

© 1985 RIBA Publications Ltd

EXAMPLE 15

Certificate of
completion of

Making good defects

Issued by: Reed & Seymore
address: 12 The Broadway, Borchester BC4 2NW

Employer: Aqua Products Ltd
address: Waterside Road, Borchester BC1 6AD

Serial no: 1

Contractor: Leavesden Barnes & Co Ltd
address: Thatchers Yard, Midley BC14 6SE

Job reference: 456

Issue date: 20 February 1991

Works: Shops & Offices
Situated at: Newbridge Street, Borchester

Contract dated: 1 June 1989

Under the terms of the above mentioned Contract,

I/We hereby certify that the defects, shrinkages and other faults specified in the schedule of defects delivered to the Contractor as an instruction have in my/our opinion been made good.

This Certificate refers to:

*Delete as appropriate

*1. The Works described in the Certificate of Practical Completion
Serial no. 1 dated 14 August 1990

*2. ~~The Works described in the Certificate of Partial Possession of a relevant part of the Work~~s
~~Serial no. dated~~

To be signed by or for the issuer named above.

Signed _____

Distribution	Original to:	Duplicate to:	Copies to:	
	1 Employer	1 Contractor	1 Quantity Surveyor	1 Services Engineer
		Nominated Sub-Contractors	1 Structural Engineer	1 File

EXAMPLE 16

Issued by: Reed & Seymore
address: 12 The Broadway
Borchester BC4 2NW

**Final
Certificate**

Employer: Aqua Products Ltd
address: Waterside Road
Borchester BC1 6AD

Serial no: 19

Contractor: Leavesden Barnes & Co Ltd
address: Thatchers Yard
Midley BC14 6SE

Job reference: 456

Issue date: 3 March 1991

Works: Shops & Offices
Situated at: Newbridge Street, Borchester

Contract dated: 1 June 1989

Original to Employer

Under the terms of the above mentioned Contract,

the Contract Sum adjusted as necessary is . . . .   £ 692,650.00

The total amount previously certified for payment to the contractor is . . £ 675,333.75

The difference between the above stated amounts is . . . . £ 17,316.25

(in words)____ Seventeen thousand three hundred &
____ sixteen pounds - 25p

*Delete as
appropriate

and is hereby certified as a balance due* to the Contractor from the
Employer/*~~to the Employer from the Contractor.~~

All amounts are exclusive of VAT

To be signed by or for
the issuer named
above.

Signed _____

*Delete as
appropriate

[1] The terms of the Contract provide that the amount shall as from the
*14th/~~21st~~ day after the date of this Certificate be a debt payable from the
one to the other subject to any amounts properly deductible by the
Employer.

This is not a Tax Invoice

Note:

[1] Payment becomes due 14 days after issue where the contract is JCT 80 or MW 80 and 21 days after issue for IFC 84.

 © 1985 RIBA Publications Ltd

EXAMPLE 17

Notification of
revision to
**Completion
date**

Issued by: Reed & Seymore
address: 12 The Broadway, Borchester BC4 2NW

Employer: Aqua Products Ltd
address: Waterside Road, Borchester BC1 6AD

Serial no: 2

Contractor: Leavesden Barnes & Co Ltd
address: Thatchers Yard, Midley BC14 6SE

Job reference: 456

Issue date: 16 June 1990

Works: Shops & Offices
Situated at: Newbridge Street, Borchester

Contract dated: 1 June 1989

Under the terms of the above mentioned Contract,

I/We give notice that the Completion Date previously fixed as

25 July   19 90

*Delete as
appropriate

*is hereby fixed later than that previously fixed;

~~*is hereby fixed earlier than that previously fixed;~~

~~*is hereby confirmed;~~

and is now

12 August   19 90

†Statement (a) is
for revisions made
prior to Practical
Completion and (b)
for revisions **after**
Practical Completion.
Delete as appropriate.

(a) †by reason of the relevant events identified in the Contractor's
notices, particulars and estimates, and/or instructions requiring as a
variation omission of work, which are set out below/overleaf;

(b) ~~†by reason of our review pursuant to clause 25.3.3.~~
The issue of amending instructions with respect to
additional parking facilities (AI 33 item 2).

Adverse weather conditions.

To be signed by or for
the issuer named
above.

Signed

Distribution

Original to:	Duplicate to:	Copies to:	
[1] Contractor	[1] Employer	[1] Quantity Surveyor	[1] Services Engineer
	[ ] Nominated Sub-Contractors	[1] Structural Engineer	[1] File

© 1985 RIBA Publications Ltd

Chapter 8

Delays and Disputes

Delay in the completion of the work is one of the most common causes of trouble encountered during the administration of building contracts. It is also one of the most justifiable causes of criticism of the building industry. It is not surprising therefore that when drafting JCT 80 the Joint Contracts Tribunal gave particular attention to the procedures to be followed when delay occurs, or is foreseen, as a result of which the contract period might have to be extended.

Two of the definitions in clause 1 of the conditions have particular significance in relation to the works being delayed. These are:

- completion date, which is defined as 'the Date for Completion as fixed and stated in the Appendix or any date fixed under either clause 25 (extension of time) or clause 31.1.3. (effect of war damage)'.
- relevant event, which is described as 'any one of the events set out in clause 25.4', which deals with the circumstances which may entitle the contractor to an extension of time.

Broadly speaking delays can be considered under four headings:

- delay caused by the contractor.
- delay caused by nominated sub-contractors and suppliers.
- delay caused by the employer or his architect.
- delay due to causes outside the control of the parties to the contract or their representatives.

Although only the last three categories are the subject of relevant events under clause 25.4 (thereby giving rise to extensions to the contract completion date), it should be noted that under clause 25.2.1.1 the contractor is required to give written notice to the architect if progress is being, or is likely to be, affected by any one of the three categories.

It should also be noted that, under clause 25.3.4, the contractor is required to use his best endeavours to prevent delay and must do all that may reasonably be required to the satisfaction of tne architect to proceed with the works. However, this does not mean the contractor is obliged to deploy additional resources or take measures which will increase his costs in order to overcome or reduce delays. Obviously, however, it is up to him to decide what measures he takes or costs he incurs in order to reduce his exposure to liquidated and ascertained damages for delays he has caused and for which he will get no extension of time.

Delays caused by the contractor (and his sub-contractors)

Coming within this heading are all those delays which can be avoided if the contractor proceeds efficiently and diligently with the running of the contract and the work itself. Under the terms of the contract the contractor is expected to take all possible steps to ensure that he has an adequate labour force on the job, that he has the specified materials on the job when they are needed and that the work is not held up by a delay on the part of sub-contractors.

If the contractor does not take all possible steps to avoid such delays, there is no provision in the contract for the granting of an extension of time on account of them. The responsibility for them is the contractor's and when they occur the contractor may render himself liable to compensate the employer by the payment of liquidated damages under clause 24 of the conditions of contract.

Ultimately, if delays caused by the contractor are of a serious nature and if the contractor fails to take reasonable steps to rectify them, then the employer is empowered under clause 27 to determine the contractor's employment under the contract.

Delays caused by nominated sub-contractors and suppliers

Although being under the general control of the contractor, nominated sub-contractors and nominated suppliers fall into a special category. Due to the element of imposition on the contractor by the employer or his architect in making the nomination it is they who must accept some responsibility in the event of the nominee's default.

Delays caused by nominated sub-contractor's and suppliers are often a source of trouble. This is partly because it is frequently difficult to establish exactly who is responsible for

'. . . delays on the part of sub-contractors . . .

such delays and partly because, where they are shown to be the sub-contractor's fault and where the contractor has done all he can to prevent them, the contractor is entitled under the conditions of contract to be granted an extension of time. The employer is thus deprived of his means of redress by way of liquidated damages under the main contract.

However, the documentation used for the nomination of sub-contractors in accordance with the terms of clause 35 of the conditions, incorporates in the Standard Form of Employer/ Nominated Sub-Contractor Agreement (Agreement NSC/2 where the basic nomination method has been used and NSC2a where the alternative method has been used) provisions safe-guarding the employer's position in these circumstances. The agreement establishes a direct contractual relationship between the employer and the nominated sub-contractor and one of its provisions is that the sub-contractor '. . . shall so perform the sub-contract that the contractor will not become entitled to an extension of time for completion of the Main Contract Works by reason of the Relevant Event in clause 25.4.7 of the Main Contract Conditions'. Thus if the sub-contractor causes delay to the main contract the employer will have an enforceable remedy against the sub-contractor, despite the general contractor's right to an extension of time.

Delays caused by the employer or his architect

Certain of the Relevant Events listed in clause 25.4 occur because of action or inaction on the part of the employer or the architect. These may be summarised as follows:

- the issue of architect's instructions in relation to discrepancies in the contract documents
- the issue of architect's instructions requiring a variation
- the issue of architect's instructions in regard to the expenditure of provisional sums except provisional sums for defined work as explained in rule 10 of SMM7
- the issue of architect's instructions with regard to the postponement of any of the work, including those dealing with antiquities under clause 34
- the issue of architect's instructions requiring completed work to be opened up or materials tested, provided such work or materials are found to be in accordance with the contract
- the failure of the architect to issue to the contractor in due time necessary instructions, drawings, details or levels
- delay on the part of the employer or others engaged by him on work not forming part of the contract
- failure by the employer to supply materials or goods which he has undertaken to supply for the works, or delay by him in supplying such materials and goods
- failure of the employer to give in due time access to or from the site or any part of the site in accordance with the contract documents
- deferment by the employer of the giving of possession of the site to the contractor, assuming clause 23.1.2 is stated in the contract appendix to apply
- by reason of the execution of work covered by approximate quantities in the contract bills where the approximate quantity does not reasonably accurately forecast the quantity of work actually required

If delay from these causes involves the contractor in direct loss or expense for which he would not otherwise be reimbursed, he will be entitled to recover such loss and expense under the provisions of clause 26, or in the case of antiquities under the provisions of clause 34.

In the event of a delay of this nature continuing the contractor may, under clause 28, determine his employment under the contract.

Delays due to causes outside the control of the parties to the contract

Clearly there are occasions when delays arise due to circumstances over which neither party to the contract, nor their respective representatives, have any control. Such delays are anticipated by clause 25 of the conditions of contract and may be summarised as follows:

- *force majeure*
- exceptionally adverse weather conditions
- loss or damage due to fire, lightning, explosion, storm, tempest, flood, bursting or overflowing of water tanks or apparatus or pipes, earthquake, aircraft and other aerial devices or articles dropped therefrom, riot and civil commotion (the specified perils as defined in clause 1.3)
- civil commotion, local combination of workmen, strike or lockout affecting any of the trades employed upon the works or any of the trades engaged in the preparation, manufacture or transportation of any goods or materials required for the works
- delay on the part of nominated sub-contractors or nominated suppliers which the contractor has taken all practicable steps to avoid or reduce
- delay due to action by the Government which restricts the availability or supply of labour, materials or fuel
- inability of the contractor for reasons beyond his control and which he could not reasonably have foreseen at the base date (as defined in clause 1.3) to secure such labour, goods or materials as are essential for properly carrying out the work
- delay by a local authority or statutory undertaker in carrying out work in pursuance of statutory obligations, or their failure to carry out such work

In the case of delays from such causes the contract provides in clause 25 for an extension of time to be granted in a similar manner as that set out above for delays caused by the employer or his architect. In such cases, however, there is no provision in the contract for the contractor to be reimbursed financially for any loss or additional expense which such delays may have caused him, it being contended that the burden of such delays shall be shared by the parties to the contract.

Delays arising from war damage must also be considered as

arising from causes outside the control of either party to the contract, but these are covered separately under clause 33.

The interpretation of certain of these clauses may give rise to some dispute and it is perhaps advisable to look at some of them more closely.

The term *force majeure* is a somewhat ill-defined legal phrase. In its broadest usage it would appear to include acts of God, war, strikes, epidemics and any direct legislative or administrative interference. Several of these matters, however, are provided for separately in the building contract and it may therefore be said in this context to cover matters of a similar nature which are not specifically dealt with elsewhere.

Delays due to weather conditions have frequently been the source of controversy in connection with delays in building work. In earlier editions of the standard form of contract the term 'exceptionally inclement weather' was used, but in JCT 80 this was changed to exceptionally adverse weather conditions'. This change was made following the unusually hot summer of 1976 when many contracts were delayed by exceptional weather which could hardly have been described as inclement.

The word 'exceptionally' is clearly the important one in this phrase and it must be considered according to the time of year and the conditions envisaged in the contract documents. Thus, if it were known at the time that a contract was let that the work was to be carried out during the winter months, and if that work is delayed by a fortnight of snow and frost during January, such a delay could not be regarded as due to exceptionally adverse weather. If, however, such work was held up by a continuous period of snow and frost, from early January until the end of March, an extension of time would clearly be justified under this clause. It is also important to consider the location of the work, as what may be regarded as exceptionally adverse weather in the south of England would not necessarily be regarded as the same in the Shetland Isles. In this respect local weather records can prove a valuable guide to the normal weather conditions one can expect to encounter in a particular area. In certain circumstances such delays may be avoided by providing in the contract for additional protection and even temporary heating arrangements.

In the case of delays by nominated sub-contractors and nominated suppliers, and delays due to the contractor being unable to obtain labour or materials at the right time, the contract conditions impose a forceful obligation on the contractor to take steps to avoid such delays occurring. Should they occur, the architect has a clear obligation to satisfy himself that the contractor has taken such steps before an extension of time is justified.

Clearly it is in everybody's interest that all possible steps to avoid delay should be taken, and when the architect's responsibilities in this connection are considered, three particular points should be borne in mind:

(1)   The architect can only grant extensions of time for delays due to the relevant events listed in clause 25 (and for war damage under clause 33), and can only certify payment of loss and expense in respect of the matters listed in clause 26 (or in relation to antiquities under clause 34). Other delays for which the employer may be responsible, cannot be dealt with by the architect.

(2)   Any delay by the employer or his architect (whether dealt with by the architect or not) may involve a breach of contract by the employer for which the contractor may have a right to claim damages at common law. Clause 26.6 envisages such delays, amongst other things, and makes it clear that the contractor's common law rights are preserved, notwithstanding any extension of time or reimbursement of loss and expense being granted under the contract.

(3)   Delays caused by the employer which cannot be dealt with under the contract (e.g. failure to give possession of the site within the period of deferment stated in the appendix) could lead to the contract period becoming 'at large', thus rendering the liquidated and ascertained damages clause inoperative.

Procedure in the event of delay

When any delay occurs, the procedure set out in clause 25 may be summarised as follows:

- Immediately it becomes reasonably apparent that the progress of the works is being delayed or is likely to be delayed, the contractor must give written notice to the architect of the material circumstances, stating the cause of delay.
- He must also identify any event which in his opinion is a Relevant Event.
- If the circumstances referred to above include reference to a

nominated sub-contractor a copy of the written notice must be sent to the sub-contractor concerned.

- In respect of each and every Relevant Event the contractor must in the notice or as soon as possible give in writing particulars of the expected effects of the Relevant Event and an estimate of the extent of the delay.
- Again where a nominated sub-contractor is involved, copies of the particulars and the estimate of the delay must be sent to the sub-contractor concerned.
- After the necessary notice has been given in the first place the contractor must keep the architect up to date with all matters arising from the Relevant Event, including any material changes in the particulars of the expected effects or the estimate of the extent of the delay.
- After he has received the notice of possible delay together with the particulars of the expected effects of the Relevant Event and the estimate of the extent of the delay, the architect must decide whether the works are likely to be delayed beyond the completion date and whether that delay is due to the Relevant Event referred to.
- If the architect is satisfied that there will be a delay and the Relevant Event has caused it, he must in writing give an extension of time by fixing such later completion date as he estimates to be fair and reasonable. The RIBA standard form of notification of a revised completion date is shown in Example 17.
- When fixing the later completion date the architect must state, also in writing, which of the Relevant Events he has taken into account and the extent, if any, to which he has had regard to any variations involving the omission of the work which may have been issued since the previous completion date had been fixed.

The contract requires that the architect must fix the new completion date not later than 12 weeks from the receipt of the written notice of delay from the contractor, or from the subsequent receipt of sufficient particulars and estimate of the extent of the delay. If the period between the receipt of the notice or the necessary particulars and the previously fixed completion date is less than 12 weeks then the architect must fix a revised completion date before the previously fixed completion date is reached.

Once the architect has granted an extension of time and fixed a later completion date he may subsequently amend that date to an earlier date if, after granting his extension of time, he has subsequently issued an instruction requiring as a variation the omission of any of the work, provided he is satisfied that it is fair and reasonable to do so. He cannot, however, fix a completion date which is earlier than the original completion date stated in the appendix to the conditions of contract.

Not later than 12 weeks after the date of practical completion the contract requires the architect to finalise the position regarding the completion date by taking one of three courses of action, namely:

- fixing a completion date later than that previously fixed if in his opinion it is fair and reasonable to do so having regard to any of the relevant events, whether upon reviewing a previous decision or otherwise, and whether or not the relevant event has been specifically notified to him
- fixing a completion date earlier than that previously fixed if in his opinion it is fair and reasonable to do so having regard to any instruction requiring as a variation the omission of any work which he may have issued after the last occasion on which he made an extension of time
- confirming to the contractor the completion date previously fixed

Whichever of these courses of action the architect takes he must put the matter in writing to the contractor.

All the conditions regarding the granting of extensions of time are subject to the proviso that the contractor shall constantly use his best endeavours to prevent delay in the progress of the works, howsoever caused, and prevent the completion of the works being delayed or further delayed beyond the completion date; and also that he shall do all that may be reasonably required to the satisfaction of the architect to proceed with the works.

Finally it should be noted that whenever the architect fixes a new completion date he must notify every nominated sub-contractor of his decision.

There is little doubt that these procedures impose upon both the contractor and the architect clear and not inconsiderable obligations whenever there is a likelihood of delay. When seeking an extension of time the contractor must be specific and the architect must deal with the matter when it arises. The intention of the contract is quite clear. Decisions on extensions of time are to be made as quickly as possible after the relevant event which is likely to cause the delay has occurred. The common practice of waiting to see what the delay actually is at the end of the contract and then arguing about whether or not an extension of time is justified has no place under the JCT 80 conditions. However, it should also be noted that the courts have taken a liberal view in relation to what may appear in the contract to be a mandatory 12 week cut-off point (see *Temloc Ltd v Errill Properties* – Court of Appeal 1987).

It must be remembered that, where optional clause 5.3.1.2 applies, the architect will be assisted by the contractual obligation on the contractor to provide and keep up to date a master programme for the execution of the works. Perhaps the greatest difficulty which may be encountered will be in deciding whether or not the notice, particulars and estimates which the contractor is required to provide are sufficient for the architect to make his decision on extending the completion date. This requires close co-operation between contractor and architect, and architects ought to be decisive in the matter and not use alleged insufficiency of particulars and estimates as an excuse for delaying the issue of extensions of time.

Reimbursement of loss and/or expense

Several of the relevant events referred to earlier, the occurrence of which may entitle the contractor to an extension of time, may also entitle him to be reimbursed for any direct loss and/or expense which he may incur as a result of the relevant event concerned materially affecting the regular progress of the works. These are the events which may be regarded as being under the control of the employer or the architect. They are listed as 'matters' in clause 26.2 and will be seen to be those events summarised earlier in this chapter as delays caused by the employer or the architect.

If the contractor considers he has incurred, or is likely to incur, direct loss and/or expense as a result of one of the matters listed he must make a written application to the architect. This application must be made as soon as it becomes apparent that the regular progress of the works is being, or will be, affected. The contractor must give the architect as much information as he can and, when required to do so, he must submit to the architect or to the quantity surveyor such details as they may reasonably require to ascertain the loss and/or expense. As soon as it is possible to do so the architect or, if he so instructs, the quantity surveyor, must ascertain the amount of loss and/or expense and the amount so ascertained must be included in the next interim certificate.

It must be borne in mind that the fact that the contractor may be granted an extension of time as a result of one of the relevant events listed in clause 25 does not necessarily mean that he is automatically entitled to some financial reimbursement. He must show that the relevant event is also a matter under clause 26 and that it has caused him loss and/or expense which he will not recover under any other provision in the contract.

Conversely it should be borne in mind that a delay may be of a type which entitles the contractor to recover loss and expense but if the delay will not affect the completion date because it has no impact on the critical path then no extension of time should be granted.

Liquidated damages

If the contractor fails to complete the works by the completion date the architect must issue a certificate to that effect and the employer may then exercise his right to receive damages for non-completion from the contractor. Whether or not the contractor is required to pay liquidated damages is now entirely at the discretion of the employer, but if he does intend to claim liquidated damages he must inform the contractor in writing before the architect issues the final certificate.

The amount of the damages will have been inserted in the appendix to the conditions of contract and is usually stated as a given sum per day or per week. It is important that this sum should be carefully and realistically calculated at the time the tender documents are being prepared. The amount should be a proper estimate of the damage the employer is likely to suffer if the work is not completed on time. It is not intended as a penalty and, should an unrealistically high figure be included, the court may well set aside the amount stated in the contract in favour of a realistic assessment of the damages actually suffered. The court, however, will not normally interfere if the amount stated is a genuine estimate of the damages.

It is sometimes argued that there is little point in putting an amount for liquidated and ascertained damages in the appendix to the contract, as the provisions of clause 24 are very seldom enforceable. There is no justification for this argument, however, provided the specific and implied terms of the contract are adhered to, particularly those in clause 25 regarding extension of time, which have already been discussed in this chapter.

Disputes and arbitration

Many disputes arise in the course of building operations, but most of these are of a minor nature and are settled fairly and amicably by those concerned. From time to time, however, matters come into dispute which cannot be easily settled and subsequently become the subjects of claims by one party to the contract against the other. It is not feasible to consider in detail what matters might give rise to disputes and how they should be dealt with, as each case must be considered on its merits. In general, it can be said that disputes and claims can be traced back to failure by one of the parties to the contract, one of the professional advisers or some other party connected with the contract to do his work efficiently, to express himself clearly or to understand the full implications of the instructions issued to or received by him.

In the event of a serious dispute arising every effort should be made to reach a fair settlement by negotiation. If, however, this proves impossible the dispute must be referred to arbitration in accordance with article 5 of the Articles of Agreement and clause 41 of the conditions. The only exceptions to this provision are disputes in relation to the statutory tax deduction scheme insofar as legislation provides other methods of settling such disputes, and disputes arising from the employer's right under the VAT agreement to challenge the amount of tax claimed by the contractor.

The procedure for referring a dispute to arbitration is quite straightforward. In the first instance the party to the contract wishing to refer the matter to arbitration must ask the other party, in writing, to concur in the appointment of an arbitrator. This appointment may be made by an agreement between the parties, or, if they cannot agree on a suitable person, the appointment will be made by the president or a vice-president of the RIBA, the RICS or the Chartered Institute of Arbitrators (the selection being set out in the appendix to the contract conditions). As the contract itself contains an agreement that matters in dispute shall be submitted to arbitration, this written request for the appointment of an arbitrator automatically refers the matter in dispute to him when appointed.

There are matters of dispute which may be dealt with by arbitration during the progress of the work. These are:

- dispute arising in connection with the appointment of a new architect or quantity surveyor in accordance with the articles of agreement
- dispute as to whether an architect's instruction is valid
- dispute as to whether a certificate has been improperly withheld or has not been properly prepared in accordance with the contract conditions
- dispute as to whether the determination of the contractor's employment after some loss under clause 22C is just and equitable
- dispute with regard to the contractor's reasonable objection to complying with an instruction
- dispute as to whether work is in accordance with the contract, and in relation to opening up for inspection
- dispute as to the withholding of consent by the contractor under the arrangements for partial possession, use or occupation by the employer
- dispute in connection with an application for extension of time
- dispute arising in connection with an outbreak of hostilities or with war damage

Arbitration on all other matters shall not be commenced until after the practical completion of the works or the determination of the contract, unless with the written consent of both parties to the contract. Any matter of dispute in connection with a contract may be referred to arbitration, including such matters as are in the ordinary way solely in the discretion of the architect.

In addition clause 41 contains joinder provisions under which separate but related disputes may be dealt with together before the same arbitrator. Thus, where there is a dispute between the employer and the contractor on the one hand, and a dispute on related issues between the contractor and a nominated or domestic sub-contractor or nominated supplier on the other hand, and where the arbitrator is a suitably qualified person to hear both disputes, the arbitrator may direct that the two references may be joined together in the arbitration proceedings. However, there is provision for the parties to opt out of the joinder provisions, provided they do so before the contract is signed and record their decision in the appendix to the conditions (clauses 41.2.1 and 41.2.2).

In 1988 the JCT issued a set of arbitration rules which are intended to be applicable to all the JCT contracts, subcontracts and warranty agreements. Only in the Agreement for Minor Building Works are the rules indicated to be optional by the inclusion of a note allowing the deletion of the relevant clause.

These rules are printed in a booklet together with comprehensive notes, which should be carefully considered. It is particularly important to realise that the rules set down a very strict timetable – much stricter than laid down in other rules or often observed in practice – and failure to comply with these time limits may have adverse consequences on the defaulting party.

Chapter 9

Insolvency

This chapter deals with corporate insolvency in the case of building contracts. No specific consideration will be given to the subject of bankruptcy, that is individual insolvency which applies to sole traders and partnerships. This is because the vast majority of contracts will be entered into by contractors who are limited liability companies. However, it is hoped that the approach of this chapter will prove helpful with reference also to individual insolvency.

It is important to have a basic understanding of insolvency before considering the relevant clauses of JCT 80.

Liquidation

The liquidation (or winding up) of a company brings its existence to an end with the object of dealing with its assets for the benefit of the creditors and members. Liquidations of insolvent companies are of two types. Firstly, there is voluntary liquidation. If members of the company realise that the company is insolvent, they can resolve to place it in liquidation. Secondly, there is compulsory liquidation where it is possible for creditors to petition the court for the liquidation of the company. If, in the opinion of the court, the company is unable to pay its debts, the court may order the winding up of the company. A provisional liquidator may be appointed by the court immediately after the presentation of the petition. This is usually the Official Receiver who is appointed to act pending the appointment of a private practitioner (usually an accountant) by the creditors.

Receivership

A receiver may be appointed by a debenture-holder, normally a bank, under an express power to do so contained in his debenture. A debenture is a document evidencing a debt generally secured by a mortgage and, upon the occurrence of a

'bankruptcy . . . . is relatively infrequent'

specific event, the debenture-holder appoints a receiver. As a result of the Insolvency Act 1986, such receivers can now be sub-divided into administrative receivers and ordinary receivers. In general, most receivers appointed under a debenture secured by a charge will be administrative receivers.

Administration

This is a procedure created by the Insolvency Act 1986. The role of the administrator is to keep the company alive as a going concern or to achieve a better realisation of the assets than would be the case in a liquidation. An application is made to the court by the company, its directors, or one or more creditors.

Voluntary Arrangements

The directors may make a proposal to the company and its creditors for a composition in satisfaction of its debts or for a scheme of arrangements of its affairs. This proposal is then put to a nominee who reports to the court. If the nominee considers that a meeting of the company and its creditors should be summoned to consider the proposal, he will give notice of dates and venues. Such a meeting will then decide whether to approve the proposed voluntary arrangement.

Whereas in all the other chapters, the practice described either happens fairly frequently or else is entirely within the control of the design team so that it can be made to happen, the procedure upon insolvency can only occur when this event takes place. Although there are very few instances of an employer becoming insolvent, liquidation of a contractor is a relatively frequent occurrence. This chapter will first deal with the procedure in the case of the contractor's liquidation. Afterwards, but more briefly, the procedure upon the liquidation of a nominated sub-contractor and the employer will be dealt with.

The procedure upon the insolvency of the contractor

The relevant contract provisions

As the understanding of the procedure is so closely tied in with the clauses of the contract, the relevant clauses of JCT 80 are quoted in full:

27.2 In the event of the Contractor becoming bankrupt or making a composition or arrangement with his creditors or having a proposal in respect of his company for a voluntary arrangement for a composition of debts or scheme of arrangement approved in accordance with the Insolvency Act 1986, or having an application made under the Insolvency Act 1986 in respect of his company to the court for the appointment of an administrator, or having a winding up order made or (except for the purposes of amalgamation or reconstruction) a resolution for voluntary winding up passed or having a provisional liquidator, receiver or manager of his business or undertaking duly appointed, or having an administrative receiver, as defined in the Insolvency Act 1986, appointed, or having possession taken, by or on behalf of the holders of any debentures secured by a floating charge, of any property comprised in or subject to the floating charge, the employment of the Contractor under this Contract shall be forthwith automatically determined but the said employment may be reinstated and continued if the Employer and the Contractor, his trustee in bankruptcy, liquidator, provisional liquidator, receiver or manager as the case may be shall so agree.

27.4 In the event of the employment of the Contractor under this Contract being determined under clauses 27.1 or 27.2 and so long as it has not been reinstated and continued, the following shall be the respective rights and duties of the Employer and the Contractor:

27.4 **.1** the Employer may employ and pay other persons to carry out and complete the Works and he or they may enter upon the Works and use all temporary buildings, plant, tools, equipment, goods and materials intended for, delivered to and placed on or adjacent to the Works, and may purchase all materials and goods necessary for the carrying out and completion of the Works;

27.4 **.2** **.1** except where the determination occurs by reason of the bankruptcy of the Contractor or of him having a winding up order made or (other than for the purposes of amalgamation or reconstruction) a resolution for voluntary winding up passed, the Contractor shall if so required by the Employer or by the Architect on behalf of the Employer within 14 days of the date of determination, assign to the Employer without payment the benefit of any agreement for the supply of materials or goods and/or for the execution of any work for the purposes of this Contract but on the terms that a supplier or sub-contractor shall be entitled to make any reasonable objection to any further assignment thereof by the Employer;

27.4 **.2** **.2** unless the exception to the operation of clause 27.4.2.1 applies the Employer may pay any supplier or sub−contractor for any materials or goods delivered or works executed for the purposes of this Contract (whether before or after the date of determination) in so far as the price thereof has not already been paid by the Contractor. The Employer's rights under clause 27.4.2 are in addition to his obligation or discretion as the case may be to pay Nominated Sub-Contractors as provided in clause 35.13.5 and payments made under clause

27.4.2 may be deducted from any sum due or to become due to the Contractor or shall be recoverable from the Contractor by the Employer as a debt;

27.4 **.3**   the Contractor shall as and when required in writing by the Architect so to do (but not before) remove from the Works any temporary buildings, plants, tools, equipment, goods and materials belonging to or hired by him. If within a reasonable time after any such requirement has been made the Contractor has not complied therewith, then the Employer may (but without being responsible for any loss or damage) remove and sell any such property of the Contractor, holding the proceeds less all costs incurred to the credit of the Contractor;

27.4 **.4**   the Contractor shall allow or pay to the Employer in the manner hereinafter appearing the amount of any direct loss and/or damage caused to the Employer by the determination. Until after completion of the Works under clause 27.4.1 the Employer shall not be bound by any provision of this Contract to make any further payment to the Contractor, but upon such completion and the verification within a reasonable time of the accounts therefor the Architect shall certify the amount of expenses properly incurred by the Employer and the amount of any direct loss and/or damage caused to the Employer by the determination and, if such amounts when added to the monies paid to the Contractor before the date of determination exceed the total amount which would have been payable on due completion in accordance with this Contract, the difference shall be a debt payable to the Employer by the Contractor; and if the said amounts when added to the said monies be less than the said total amount, the difference shall be a debt payable by the Employer to the Contractor.

Procedure during contract

The design team will usually be aware of the contractor's financial difficulties before any formal steps are taken in relation to insolvency. This may be because of rumours circulating around the site or because of the rapid depletion of the contractor's resources, such as his workforce. The fact that the contractor is, or appears to be, in financial difficulties does not necessarily mean that he is insolvent. It is important, therefore, that the architect and the quantity surveyor should keep the matter strictly confidential because, apart from the possibility of libel or slander actions, the spread of this information could quickly drive the contractor into more serious difficulties or even liquidation.

However, it would be a sensible precaution for the quantity surveyor to check that the contractor has paid amounts included for nominated sub−contractors in the previous certificate. Where the contractor has not paid these sums, the employer has in some circumstances a right to pay the sub-contractors direct and to deduct the same from any sums due to the contractor. Clause 35.13 of JCT 80 deals with this situation. However, it is important to note that this right shall cease to have effect absolutely if the contractor has a petition presented against him for winding up or the shareholders pass a valid resolution for voluntary liquidation: Clause 35.13.5.4.4. The footnote to this clause states that, in the case of a contractor subject to bankruptcy law and not the law relating to the insolvency of a company, this clause 'will require amendment to refer to the events on the happening of which bankruptcy occurs'.

Immediate procedure

Clause 27.2 provides for the determination of the contractor's employment in the event of his insolvency. If any of the events in this clause constituting insolvency occur, then the employment of the contractor is said to come to an end automatically. There is no need for any formal letter determining the employment of the contractor; this is made abundantly clear by the use of the words 'forthwith automatically' in the contract. An employer should, however, take immediate steps to tell the contractor, as well as his liquidator/provisional liquidator/receiver, that he regards the employment of the contractor as determined in accordance with the clause. This should be sufficient to avoid any suggestion that by carrying on as if the contractor's employment had not been determined, it has been 'reinstated and continued' by agreement as contemplated in the clause.

It is also important to note that it is only the employment of the contractor which is determined, not the contract itself. The contract goes on and the employer still has various rights under it.

It is desirable that as soon as the liquidation becomes known, the site should be closed and no materials or plant allowed to leave it. There will be various demands made by suppliers and sub-contractors for the return of materials. It is important, therefore, to consider the position regarding ownership of materials.

It is a well-established rule that property to all materials and fittings, once incorporated in or affixed to a building, will pass to the owner of the freehold: see *Appleby* v *Myers* (1867) LR 2 CP 651. The employer may well not be the freeholder but this should make little difference, in practice, to this rule.

So far as the ownership in unfixed materials and goods is concerned, one has to look at clauses 16, 30.2 and 30.3 of JCT 80.

Under clause 16.1, the property in unfixed materials and goods passes to the employer where their value, in accordance with clause 30.2, has been included in a certificate and paid for by the employer. Clause 14(1) of JCT 63 (the equivalent to clause 16.1 of JCT 80) was considered in the case of *Dawber Williamson Roofing Limited* v *Humberside County Council* (1979) 14 BLR 70. The plaintiffs were sub-contractors. They delivered roofing slates to the site for the purpose of subsequent fixing. The value of the slates was included in a certificate under the main contract, and the certificate was paid. The main contractor did not pay the plaintiffs and then went into liquidation and their employment was determined. The defendants, the employer, refused to allow the plaintiffs to remove the slates on the basis that, by virtue of clause 14(1), the property had passed to them. The court held that ownership in the slates had not passed from the sub-contractor to the main contractor and accordingly had not passed to the employer who was therefore liable in damages to the plaintiffs. This was because the plaintiffs, as sub-contractors, were not parties to the main contract and therefore could not be bound by anything in it.

As a result, the Joint Contracts Tribunal issued Amendment 1 (1984) to JCT 80 (which is marked 'Not for Use in Scotland'). Clause 19.4.2.2 provides that 'where in accordance with clause 30.2 of the Main Contract Conditions, the value of any such materials or goods shall have been included in any Interim Certificate under which the amount properly due to the Contractor shall have been discharged by the Employer in favour of the Contractor, such materials or goods shall be and become the property of the Employer and the Sub-Contractor shall not deny that such materials or goods are and have become the property of the Employer'. Clause 21.4.5 was introduced in August 1984 as Amendment Number 1 to the DOM/1 conditions. There are similar provisions in Amendment Number 2 (1984) clause 21.2.4.2 of NSC 4 for nominated sub-contractors.

Clause 19.4.2.3 of JCT 80 also provides that every domestic and nominated sub-contract should include the term that, if the main contractor pays the sub-contractor for unfixed materials and goods before the value, in accordance with clause 30.2, has been included in any interim certificate, such materials and goods shall upon such payment by the main contractor be and become the property of the main contractor. See clause 21.4.3 of NSC 4 and clause 21.4.5.3 of DOM/1.

Clause 16.2 of JCT 80 provides that where the value of any materials or goods intended for the works and stored off site, in accordance with clause 30.3, has been included in any interim certificate under which the amount properly due to the contractor has been paid by the employer, such materials and goods shall become the property of the employer. The contractor shall not thereafter be entitled to remove such materials and goods from the premises where they are stored. The architect is bound to certify the total value of the materials and goods delivered to or adjacent to the works, provided they have been properly and not prematurely delivered and are adequately protected (clause 30.2.1.2). He has a discretion whether to certify for materials and goods before delivery (clause 30.3).

The architect's duty to certify for materials and goods under clause 30.2.1.2 may cause difficulties if, for example, the sub-contractor had no title to goods in the first place. This may be because his own supplier has delivered the goods to him subject to a Romalpa (or retention of title) clause. This clause states, in its most basic form, that title will be retained by the supplier in goods and materials unless and until they have been paid for.

The law on Romalpa clauses (so called because of the case of *Aluminium Industrie Vaassen BV* v *Romalpa Aluminium Limited* [1976] 2 All ER 552) is complex. The concept has been progressed by subsequent case-law, where, for example, goods sold under a contract including a Romalpa clause are no longer in their original state. It would not be possible to go into detail about this area of the law in this chapter and readers are advised to consult standard works on contract. It is doubtful that the amendments made to JCT 80, NSC 4 and DOM/1 which are referred to above will have the effect of validly transferring property in goods and materials to the employer/main contractor if the sub-contractor was not the owner in the first place because, for example, his own supplier has retained title to them until payment.

However, the supplier himself may also be divested of his title to goods by virtue of the operation of section 25 of the Sale of Goods Act 1979. This section provides in effect that the contractor can, by reselling the goods to a *bona fide* purchaser as a 'buyer in possession', pass a good title even though the contractor himself lacked legal title at the time. The operation of this statutory provision in the context of title retention was illustrated in the Scottish decision of *Archivent* v *Strathclyde Regional Council* (1984) 27 BLR 98. The Scottish decision is not, of course, binding on English courts but the principle at issue was broadly similar to that involved in the English case of *Four Point Garage Limited* v *Carter* [1985] 3 All ER 12. In the Archivent case, ventilators were supplied to the main contractors under a Romalpa clause. After delivery to the site, the value of the ventilators was included in an interim certificate issued under the main contract (JCT 63) and the contractors received payment for them from the employers. The contractors went into receivership leaving their debt to the suppliers unpaid.

The Scottish Court of Session held that title had passed to the employers under section 25 of the Sale of Goods Act 1979. The conditions set out in that section had been fulfilled: the contractors had received possession of the ventilators with the consent of the original owners (the suppliers); there had been a disposition by the contractors to the Council when the ventilators had been measured and valued at the site; the Council had taken the goods under the disposition in good faith and without any notice of the suppliers' original title; and in disposing of the ventilators the contractors had acted in the ordinary course of business as mercantile agents with an ostensible authority to make the disposition.

In summary, there is no easy method of determining title in the case of unfixed materials and goods. In practice, it is probably safest to assume that all materials on the site are in the ownership of the employer and to allow no one onto the site under any pretext whatsoever. It is then up to the liquidator or to a sub-contractor or supplier to prove that ownership in particular materials has not vested in the employer. However, there is always a risk that if title to materials and goods has not passed to the employer, the supplier or sub-contractor may seek an injunction for delivery up of the goods and materials in question or damages for their detention or conversion. In the event of a possibility of such a dispute, legal advice must be sought.

If a clerk of works is employed, he may be able to act as watchman during the day. However, it will be prudent for the employer to engage a watchman whenever the clerk of works is not on the site. The danger of losing materials from a building site is always high, but it is much higher in these particular circumstances.

The quantity surveyor will almost certainly, in any subsequent work for completing the job, require a list of unfixed materials on site. It is therefore a good idea for him to make this list as soon as possible. He and the architect should try to make immediate arrangements with the employer for getting all those materials which are valuable and which can be moved under lock and key. It may well be that some or all of this type of material is already locked up, in which case the locks should be changed.

Under clause 27.4.1 of JCT 80, the employer may employ and pay other persons to carry out and complete the works and he or they may use all temporary buildings, plants, tools, etc. It is important to remember that these rights of the employer (when based upon a determinaion by insolvency) are dependent upon, firstly, a valid right in law to use such buildings, plant, tools, etc. Such a clause, it is often argued, is void against the liquidator as a matter of law. This is because the liquidator, upon his appointment, takes into his custody and control all the property of the company. There is a second, more practical, difficulty. These rights cannot bind persons who are not parties to the contract. Many types of plant are hired by the contractor or owned by a subsidiary company. No provision in the contract between the employer and the contractor can, therefore, be of benefit against the true owner. If necessary, therefore, the owners of hired items may well have to be contacted in order to enter into a new agreement.

Completion of contract

It will be necessary to arrange a meeting with all concerned to consider the best method of completing the contract. The employer, his professional advisers and all consultants should be at this meeting. When the liquidator is appointed, however, he should be kept informed of the decisions which have been made.

If a bond is coupled with the contract, then the bond-holder should be kept informed. It is suggested that the employer should check the tems of the bond immediately on becoming aware that the contractor is in financial difficulties. It should be ensured that all requirements under the bond (for example, as to notice) are complied with.

The liquidator can, subject to there being a valid determination of the contract by the employer, disclaim any onerous or unprofitable contract. It is no longer necessary for the liquidator to obtain the leave of the court before exercising his power of disclaimer. The disclaimer terminates the rights and liabilities of the company in the property disclaimed but does not, over and above this, affect the rights and liabilities of any other person. An interested party, so that it should not be kept on tenterhooks, may require the liquidator in writing to decide whether he intends to disclaim. If the liquidator has within 28 days failed to give a notice of disclaimer, then his right to do so is lost.

Alternatively, under the conditions of contract, the employer and the liquidator/provisional liquidator/receiver may agree to reinstate and continue the works. It is noticeable that, whilst the JCT has included references to an administrative receiver and an administrator within the body of clause 27.2, any reference to employment being reinstated and continued by agreement between the employer and the administrative receiver/administrator has been omitted. The third, and perhaps the most usual, of course, is for the employer to complete the contract.

The procedure to be adopted for the completion of the contract in the last mentioned case will depend upon the extent of the work to be completed:

(1) *If the contract was only just started*, then it would be reasonable to go out to tender again asking the same type of contractor to tender for the completion of the works as in the original tender list. In this case, the quantity surveyor could deal with the tendering documents by preparing a short bill of the work already carried out which would be an omission bill and that, coupled with the original bill, could form a bill of quantities of the work to be completed.

(2) *If the work is well in hand but far from complete*, it may be best to get competitive tenders for completing, though some-times it may be advantageous to negotiate with a single contractor a suitable contract for completion. The employer might be in difficulties in later dealings with the bond-

holder or liquidator if he has failed to convince them that he has taken due care to complete the work in a reasonably economical way. It would be as well if the employer could show that negotiating a contract is more economical than going to competitive tender again. Provided the negotiations are reasonable this may not be too difficult to show if time has been saved thereby.

The quantity surveyor will have to use his own judgement in these circumstances, as to whether it is quicker for him to produce a new document for the completion of the work, or whether it is still better to use the existing bills with an omission bill for work which has already been carried out.

(3) *If the work is substantially complete*, what remains to be done may very often be in the nature of jobbing work. In these circumstances, a completely different type of builder may be the best one to complete. Furthermore, it may be better to complete on a prime cost contract having a fixed fee. It will be up to the quantity surveyor to weigh up the job with particular reference to the sub-contractor's work still to be done, and to make recommendations on the most suitable contract procedure for completion. If the client has a direct labour force, it may be convenient for him to complete it in this manner, especially when one bears in mind that the existing sub-contractor may well be available to carry on.

One item that will be essential in the bills for completion will be a provisional sum to cover making good of defects left by the contractor in liquidation. Where these are actually known they can of course be measured, but it is necessary to make sure that this is covered as there will probably be some defects which do not make themselves apparent until the new contractor is on the site.

Most of the work in getting out the new contract documents will necessarily fall on the quantity surveyor. It will probably not be necessary for the architect to do anything other than re-issue his existing drawings as it will be obvious to the new contractor that he is only to execute that work which has not already been done. For the purpose of establishing a price, however, the quantity surveyor will have to prepare documents in one of the ways mentioned above and it is essential that he do this as quickly as possible.

Nominated sub-contractors

Under clause 31 of NSC/4, if the employment of the contractor is determined under clause 27 of the main contract, the employment of the sub-contractor is also determined. There will very likely be a considerable number of sub-contractors to deal with and they will fall into two groups:

(1) *Those who have not yet started the works*
As sub-contracts should be drawn up in terms similar to the main contract, there should be no difficulty for those who have not started. They will be invited to enter into a similar sub-contract with the new contractor when the new contract is placed.
(2) *Those who have partially completed the works involved*
Where sub-contractors have partially executed their work, it will probably be best for the quantity surveyor to negotiate with each in turn a price for completing. The quantity surveyor must take care, however, only to measure and assess the value of the work to be completed and not take any account of the fact that the sub-contractor may not have been paid for all the work he has already done.
    The quantity surveyor, in negotiating for the completion of the sub-contract, should bear these matters in mind because it may well be that the sub-contractor will not agree to complete unless he is paid in full for the work done before the winding up was commenced. The surveyor's attention may, therefore, have to be directed to what it will cost to complete the work by another sub-contractor. This may mean the negotiation for the completion of the sub-contract work will be on higher rates than the original quotation. There will be an obvious advantage, therefore, to the employer if the same sub-contractor can be persuaded to complete. Clauses 35.13.2 to 35.13.5 of JCT 80 cover the position regarding direct payment to nominated sub-contractors. It is important to note, however, that the right of the employer to pay a nominated sub-contractor direct will cease if the contractor company has had a petition presented against it for winding up, or the shareholders pass a valid resolution for voluntary liquidation: clause 35.13.5.4.4.
    The cessation of the employer's right to make direct payments to a nominated sub-contractor is thought to be the result of a decision of the House of Lords in the case of

British Eagle International Airlines v *Compagnie Nationale Air France* [1975] 1 WLR 758. This appeared to overrule the case of *Re Tout and Finch* [1954] 1 All ER 127 which dealt specifically with the employer's right under the JCT contract to pay sub-contractors direct. The British Eagle case threw doubts on certain direct payments where insolvency of the main contractor occurs, for such payments would be in danger of being declared invalid at the instigation of the liquidator. This may then result in the employer having to pay twice, once to the sub-contractor and once to the liquidator.

Clause 7.2 of NSC/2 and clause 6.2 of NSC/2a provide that if, after paying any amount to the sub-contractor under clause 35.13.5.3 of JCT 80, the employer produces reasonable proof that there was in existence at the time of such payment a petition or resolution to which clause 35.13.5.4.4 of JCT 80 refers, the sub-contractor *shall* repay on demand such amount.

Bond

The contractor may be required to take out a bond for the due completion of the work and this is usually in the sum of 10 per cent of the contract sum. Subject to its terms, the bond may be used where the employer is put to additional expense through the contractor's liquidation. Bond-holders are usually insurance companies specialising in this business but may sometimes be banks. Although they may be liable for paying the employer such extra money as is necessary to complete the job, they have no control over the way the contract is completed. On the other hand, it is up to the employer to complete it without any undue extravagance and it is just as well, to avoid future repercussions, to keep the bond-holder informed of what is happening.

Final account

Clause 27.4.4 of JCT 80 provides that '...the Contractor shall allow or pay to the Employer in the manner hereinafter appearing the amount of any direct loss and/or damage caused to the Employer by the determination.' This clause deals with the respective rights and duties of the contractor so long as the latter's employment has not been reinstated and continued. Strictly speaking, this could mean that the contractor would be entitled to payment but, in practice, this is seldom likely to be the case. The employer is entitled to the expenses of completing the works and to any damages which he has suffered by reason of the determination. It appears that this would include any extra fees and expenses which are incurred and the costs of delay.

For example, the architect may well have had additional visits to the site, additional prints of drawings may have been required and so on. The quantity surveyor would have had a great deal of extra measurement as well as the complication of sorting out the final accounts. The contractor has to allow or pay expenses properly incurred by the employer and the amount of any direct loss and/or damage caused to the employer by the determination, subject to receiving a credit for what he would have been paid had the contract not been determined. Thus, two final accounts will have to be prepared:

(1) The first, or notional, final account will be the amount which the contract would have cost had the first contractor completed in the ordinary way. This means that not only must all variations ordered before the winding up be measured and priced at the rates relevant to the first contract, but also that all variations ordered on the completion contract must be included and priced at the rates which would have pertained had the first contract continued.

(2) The second final account will be a normal final account for the completion contract and will only include those variations ordered during the completion contract. They will be priced at the relevant prices and rates for the completion contract, which may well be different from the first contract.

One point that needs to be watched is that if a completion contractor has to make good any defects which were caused by the contractor in liquidation and these are ordered as a variation on the completion contract, they should not show as a variation on the first final account because they would not have been included in that final account had the first contractor completed normally.

A point arises as to the extent of the variations ordered on the completion contract which can be taken into account on the notional final account. Normal variations will go into account but where extras are exceptionally large or involve a change in the character of the job, the liquidator may have reasonable grounds for objection.

An example of a statement showing the financial position of all the parties at completion is illustrated in the cases below.

(1) It has been assumed that the contract was able to be completed without too much trouble and there is still something to be paid to the contractor in liquidation. It must be remembered that when the winding up was commenced there was probably considerable money in the retention fund and also, unless a certificate had been issued immediately before, there would be some money for work done but not yet certified. Therefore, it is assumed that the extra cost of completing the contract was less than the monies outstanding to the contractor at the time of their liquidation and therefore there is finally a debt due from the employer to the liquidator.

(2) In this illustration the extra cost of completing the work was considerably greater than the money outstanding at the time of the liquidation. A retention sum would have been relatively small as the contract was only one-third on. If there had been a bond for 10 per cent of the contract sum, then the employer would, depending upon the terms of the bond, first recover from the liquidator the amount in the pound which was being paid out, and then get up to a maximum of £20 000 from the bond-holders. If the final dividend were 10p in the £ he would receive £3 030 from the liquidator, leaving £27 270 of which the bond-holder would have to pay £20 000, reducing the employer's ultimate loss to £7 270. If, on the other hand, the final dividend were 80p in the pound he would receive £24 240 from the liquidator and the bond-holder would provide the balance of £6 060.

The agreement of the second final account should create no difficulty because there is a contractor's organisation with which to work. However, the first final account will probably have to be done by the quantity surveyor alone and there will be very little opportunity for the contractor who has gone into liquidation to check and agree it as his staff will probably not be available. The liquidator should be informed of the way the final account is being prepared and sent a copy at completion for his agreement.

CASE (1) – *Resulting in debt from employer to contractor*

	£	£
Amount of original contract (with company in liquidation)		200,000
Additions (whether ordered with company in liquidation or completion contractor, but priced at rates in original contract)....................................		10,000
		210,000
Omissions (same rules as for additions)...		7,500
Amount of notional final account if original contractor had completed ..		202,500
Amount of completion contract..	30,000	
Additions (for completion contract only and priced at relevant rates in the completion contract)..	2,500	
	32,500	
Omissions (ditto)..	2,000	
Amount of final account of completion contract..	30,500	
Additional professional fees incurred	1,500	
Amount certified and paid to original contractor before liquidation........	169,000	201,000
Debt payable by employer to liquidator or receiver		£ 1,500

CASE (2) – *Resulting in debt from contractor to employer*

	£	£
Amount of original contract (with company in liquidation)		200,000
Additions (whether ordered with company in liquidation or completion contractor, but priced at rates in original contract)		10,000
		210,000
Omissions (same rules as for additions)...		7,500
Amount of notional final account if original contractor had completed ..		202,500
Amount of completion contract..	150,000	
Additions (for completion of contract only and priced at relevant rates in the completion contract)	10,500	
	160,500	
Omissions (ditto)..	7,700	
Amount of final account of completion contract..	152,800	
Additional professional fees incurred	7,500	
Amount certified and paid to original contractor before liquidation	72,500	232,800
Debt payable by liquidator or receiver to employer and partly by bondholder		£ 30,300

Retention Monies

Under clause 30.5 of JCT 80, the employer's interest in the retention is fiduciary as trustee for the contractor and for any nominated sub-contractor. This is to protect a contractor and sub-contractor in the event of the employer's insolvency. However, if the contractor becomes insolvent, it may still be open to the liquidator of the contractor, in a situation where the employer withholds the whole of the retention monies from the contractor, to proceed against the employer as trustee for the sub-contractor. Thus, the employer may not be entitled to have recourse to the retention money held for the benefit of the nominated sub-contractor. Retention monies, if the contractor or nominated sub-contractor request, will have to be placed by the employer in a separate bank account, appropriately designated (see clause 30.5.3).

The procedure upon the insolvency of a nominated sub-contractor

Clause 35.24.7 of JCT 80 states that where the nominated sub-contractor becomes insolvent (that is, where clause 35.24.2 applies), the architect shall make such further nomination of a sub-contractor in accordance with clause 35 as may be necessary to supply and fix materials or goods or to execute the work and make good or re-supply or re-execute any defective work or materials. This confirms the decision made in the case of *Bickerton v N.W. Metropolitan Regional Hospital Board* [1970] 1 All ER 1039. Under clause 35.24.10 of JCT 80 (amendment 5: 1988), the architect is under an obligation to make the further nomination of a sub-contractor within a reasonable time, having regard to all the circumstances, after the nominated sub-contractor has become insolvent (as set out in clause 35.24.2). The amount payable to the nominated sub-contractor under the sub-contract resulting from such further nomination is to be included in interim certificates and added to the overall cost of the works.

It seems clear, from the decision of the House of Lords in the case of *Percy Bilton Limited v GLC* [1982] 1 WLR 794 (which is a decision on the terms of a JCT 63 contract), that delay inevitably occurring due to the departure or (dropping out) of a nominated sub-contractor does not entitle the main contractor to any extension of time under clause 23(g) of the main contract (the equivalent clause to clause 25.4.7 of JCT 80). However, the employer, through his architect, is under a duty to renominate or replace within a reasonable time. This is now enshrined in clause 35.24.10. Insofar as any period of delay is attributable to failure to renominate in due time, it would appear that this would now come under clause 25.4.5.

Therefore, under JCT 80 an architect has power to extend the contractor's time for completion if a nominated sub-contractor goes into liquidation and there is a subsequent undue delay in renomination. There is no corresponding ground for loss and expense in clause 26 as a result of an extension under clause 25.4.5. It may well be, therefore, that the contractor will make a claim for an extension of time under clause 25.4.6, (that is, that he did not receive in due time a necessary instruction from the architect for which he specifically applied in writing) so that he can seek loss and expense under clause 26.2.1 of JCT 80.

The principles by which it may be decided whether the employer has renominated within a reasonable time were discussed by His Honour Judge Smout QC at first instance in the important case of *Rhuddlan Borough Council* v *Fairclough Building Limited* (1985) 3 Con LR 38. His Honour's judgment, as well as the Court of Appeal's judgment, should be read in detail. The Court of Appeal dealt with a number of important issues and decided, amongst other matters:

- that a main contractor can refuse to accept the renomination of a substitute sub-contractor who does not offer to complete his part of the work within the overall period for the contract
- following the Bickerton case, that on the true construction of clause 27 of JCT 63, the main contractor is neither bound nor entitled to do any of the sub-contract work himself (see clause 19.5.2 of JCT 80). The employer must, therefore, order its omission or issue a variation order and pay the contractor for it, or negotiate a new sub-contract covering remedial and uncompleted work.

Finally, where a receiver/administrative receiver/ administrator is appointed, the architect may postpone the duty to renominate if there are reasonable grounds for supposing that the receiver/administrative receiver/administrator is prepared to continue work on the sub-contract 'in a way that will not prejudice the interests of the Employer, the Contractor or any Sub-Contractor whether Nominated or Domestic engaged or to be engaged in connection with the Works.' (clause 35.24.7 JCT 80).

The procedure upon the insolvency of the employer

By clause 28.1.4 of JCT 80, if the employer becomes insolvent (in the same manner as is described for the contractor in clause 27.2 JCT 80), the contractor may determine his own employment. The contract still remains. The contractor may forthwith determine his employment under the contract by notice by registered post or recorded delivery to the employer or architect; such notice must not be given 'unreasonably or vexatiously'.

As far as the contract is concerned, it is unlikely to proceed. After taking into account amounts previously paid under the contract, the contractor shall be paid by the employer:

28.2. 2. 1. the total value of the work completed at the date of determination, such value to be computed as if it were a valuation in respect of the amounts to be stated as due in an Interim Certificate issued under clause 30.1 but after taking account of any amounts referred to in clauses 28.2.2.3 to .6;

.2 .2 the total value of work begun and executed but not completed at the date of determination, the value being ascertained in accordance with clause 13.5 as if such work were a Variation required by the Architect under clause 13.2 but after taking account of any amounts referred to in clauses 28.2.2.3 to .6;

.2 .3 any sum ascertained in respect of direct loss and/ or expense under clauses 26 and 34.3 (whether ascertained before or after the date of determination);

.2 .4 the cost of materials or goods properly ordered for the Works for which the Contractor shall have paid or for which the Contractor is legally bound to pay, and on such payment by the Employer any materials or goods so paid for shall become the property of the Employer;

.2 .5 the reasonable cost of removal under clause 28.2.1;

.2 .6 any direct loss and/or damage caused to the Contractor or to any Nominated Sub-Contractor by the determination.

The employment of the architect and quantity surveyor does not necessarily end with the employer's liquidation. The terms of the architect's and the quantity surveyor's appointment should be checked. In any event, if the architect and quantity surveyor are to continue with their services, they should seek undertakings from the liquidator that they will be paid for their services. If such undertakings are forthcoming, they can then proceed with completing their respective duties in the winding down of the contract, the settling of accounts and claims and so on.

Index

access to site, 25
accounts, final *see* final account
Agreement for Minor Building Works, 108
Agrément Certificate, 37
Aluminium Industrie Vaasen BV v Romalpa Aluminium Ltd (1976), 117
Arbitration Acts 1950 and 1979, 15
arbitration, 82
  matters dealt with, 107
  procedure for referring dispute, 107
  reference to, 7
architect,
  certificate, final, 67, 90
    example of, 94
    interim, 65–81
    of practical completion, 66
    revision to completion date, 95
  final account, 90
  inspections, routine site, 34
  instructions, 47, 49
    authority to issue, 47, 49
    direct loss and/or expense, 54
    distribution of, 50
    emanating from consultants, 50
    expenditure of provisional sums, 52
    nomination of sub-contractors, 55
    nomination of suppliers, 59
    oral/verbal, confirmation of, 48
    procedures for issue, 49
    procedures to effect nomination, 56
    site book, 49
    standard form, 62
    valuing variations, 52–4
    variations, 47–64
  insurance, responsibility to check, 19
  meetings,
    agenda, 24, 29
    example, 31
    initial site, 31
    subsequent site, 29
  progress photographs, 36
  records and reports, 35
  resident, 34
  responsibilities in respect of delays, 102
  rights, duties and responsibilities, 7–9
  samples and testing, 34, 36, 37
  site, duties 33–46
    inspections, 34, 35
  site, visits, 29, 33
  see also meetings, site
Architect's Appointment, RIBA, 33
Archivent v *Strathclyde Regional Council (1984)*, 118
Articles of Agreement, 14, 107

bankruptcy, *see* liquidation, insolvency
Bickerton v *N.W. Metropolitan Regional Hospital Board (1970)*, 127
bills of quantities, 12
  errors in, 12
  for completion after liquidation, 120
  provisional and prime cost sums, 27
British Standards Institution, 37
British Standard Specifications, 37
Building Act 1984, 34
building team, 1–11

cash flow, 65
certificates,
  arbitration, 66
  architect, 66
  completion of making good defects, 66, 93
  contractor's payment obligation to nominated sub-contractors, 65
  final, 66, 82, 90–1
    duty of architect to issue, 90
    example of, 94
  interim, example of, 79
  nominated sub-contractors, 72, 74–5
  Notification to Nominated Sub-Contractor, 81
  obligations of,
    employer, 65
    parties to the contract, 65
    quantity surveyor, 68
  payment to the contractor, 65
  practical completion, 73, 92
  retention, 71–4
    early release of, 71

proof of payment by contractor, 74

sectional completion supplement, 71

statement of retention and nominated sub-contractor's values, 78

valuation and certificate forms, 76, 77

Value Added Tax, 74–5

clerk of works,
  diary, 36
  duties, 10
  site directions, 49
  standard forms of report, 35, 45–6
  works progress report, 45–6

Code of Practice for Single Stage Selective Tendering 1977, 13

Codes of Practice, 37

communications, 11
  lines of, 28

completion, defects and the final account, 82–95

completion of contract, 119–21

completion of works, 87

conditions of contract, 14

construction insurance, 19

consultants,
  duty in respect of final account, 90
  services, 1
  structural, 1

contingency sum, 60, 61

contract,
  completion date, 26
  completion of making good defects, 82, 93
  defects liability period, 83
  delays arising from war damage, 100
  delays due to weather conditions, 101
  disputes, referral to arbitration, 52
  determination of contractor's employment, 97
  documents, 14–16, 25
  drawings, 17
  extension of time, 14, 97, 98
  failure to complete works by completion date, 106
  liquidated damages, 97
  nominations, 27
    'with documents' option, 57
    'without documents' option, 57
  period of final measurement, 89
  practical completion, 82

by a nominated sub-contractor, 58
  certificate of, 92
  procedures in the event of delay, 90, 102
    notification of revision to completion date, 95
  relevant events, 90
  sectional completion, 82
  sum, 59
  under hand, 16
  under seal, 16
  variations, 53

contractor,
  and arbitration, 52
  and architect's instructions, 52
  and interim certificates, 69
  authority to act, 34
  common law rights, 102
  delays, responsibility for, 96
  determination of the contract, 99
  direct loss or expense, recovery 99
  in liquidation, 124
  insurance of the works, 14
    relieved of obligations, 83,
  insurance requirements, 18
  master programme, 26
  nominated sub-contractors, 27, 52–5
    proof of payment, 56, 91, 92
  programme, 18–26
  rights, duties and liabilities, 4–7

Court of Appeal, 128

damages, liquidated and ascertained, 83, 85, 97, 106

Dawber Williamson roofing v Humberside County Council (1979), 116

dayworks, 55

defective work, 82

defects and making good, 86–7
  schedule of defects, 86

defects liability period, 66, 83, 84, 86, 90

delays and disputes, 96–101
  caused by contractor (and sub-/contractors), 97–8
  caused by employer or architect, 99
  due to causes outside the control of parties to contract, 100–102
  procedure, 102–5

design team, 1, 12, 17, 26

disputes and arbitration, 107–9

district surveyor, 61

domestic sub-contract (DOM/1), 117

drawings showing building as constructed, 85
see also record drawings

employer
insurance requirements, 19
obligation after receipt of architect's certificate, 65, 66
rights, duties and liabilities, 3–4
Employer's Liability (Compulsory Insurance) Act 1969, 18
engineer
consulting, 2
structural, 37

final account, 66, 83, 87–90, 123
contractual duty to complete, 90
for completion contract, 124
notional, 124
responsibilities of contractor, 87
responsibilities of quantity surveyor, 87
final certificate *see* certificates
final measurement, period of, 89
Finance Act 1985, 16
fire officer, 61
fluctuations, 68
force majeure, 100, 101
foreman, general, 34
Form of Building Contract
with Contractor's Design (JCT81), 15
Form of Nomination (NSC/3), 56
Form of Tender and Agreement (NSC/1), 56
Four Point Garage Limited v *Carter (1985)*, 118

handover meeting, 85
Health and Safety at Work Act 1974, 25
House of Lords, 127

insolvency, 119–130
administration, 111
administrator, 128
appointment of, 111
Bond, 123
completion of contract after, 120–21
of contractor, 112–119
extra fees and expenses, 124
final account, 123
financial position of parties at completion, 125

liquidator, 112, 115, 119, 120, 125
appointment of, 110
manager, 112, 113, 115
nominated sub-contractor, 122–127
payment to sub-contractors, 122
procedure, 112–130
during contract, 114
immediate, 115
receiver, 112, 113, 115
appointment of, 112
relevant contract clauses (JCT80), 112–14
voluntary arrangements, 111
Institute of Clerks of Works, 35
insurances, 18–19

JCT Standard Form of Building Contract (JCT80), 2, 14, 15, 16, 26, 47, 52, 55, 56, 72, 73, 83, 90, 101, 105, 110, 112, 117, 119, 122, 123, 127, 128, 129
appendix, 15
Employer/Nominated Sub-Contractor Agreement (NSC/2) or (NSC/2a), 23, 56, 72
Form of Tender for nominated suppliers, 59
supplements to, 15, 16, 83
JCT Standard Form Private Edition with Quantities (1980), 2
JCT63 contract, 118, 127
Joint Contracts Tribunal, 90, 116

Latent Damage Act, 16
Limitation Act, 16
liquidated and ascertained damages, 19
liquidated damages and bonuses, 106
liquidation, *see* insolvency

maintenance manuals, 87
master programme, 14, 17
meetings, site,
agenda, 29, 31
minutes, 29, 32
progress photographs, 36
records and reports, 35
samples and testing, 36–7
standard check list, 35, 37–44

National Joint Consultative Committee for Building (NJCC), 17
nominated sub-contractor(s), *see* sub-contractors

Official Receiver, 110

parent company guarantee, 17, 21
partial possession, 84–5
  of employer, 15
Percy Bilton Limited v *GLC 1982*, 127
performance bond, 16–17, 21
person in charge, 10
placing the contract, 12–23
possession of building, 85
practical completion, 82–3
Pre-Contract Practice, 12, 27, 47, 56, 59
prime cost sum (p.c. sum) 59, 61
procedure for maintenance or repair,
  85
programme, 25, 26
  see also master programme
progress meetings, 24–32
provisional sum, 53, 61

quantity surveyor, *see* surveyor, quantity

receivership, 110
record drawings, 87
records and reports, 35
reimbursement of loss and/or
    expense,
  entitlement of contractor to, 105–6
relevant event, 96, 99, 102, 103, 104,
    105, 106
retention, 71, 82
  early release of, 71
  release of half, 83, 84
Rhuddlan Borough Council v *Fairclough
    Building Limited (1985)*, 128
Romalpa clauses, the law of, 117

Sale of Goods Act 1979, 118
schedule of defects *see* defects and
    making good
Scottish Court of Session, 118
sectional completion, 83–4

Sectional Completion Supplement, 10,
    15, 16
site
  access to, 25
  agent 1, 34
  dates for possession of, 14
  duties, 33–46
standard form, 1980, *see* JCT80
Standard Method of Measurement
    (SMM7), 51, 52
structural engineer, *see* engineer,
    structural
sub-contracts, 27–8
sub-contractors, 1, 55, 58, 67
  in liquidation, 58
  nomination of, 55, 59
  standard form, 56, 57
suppliers, nomination of, 59
surveyor, quantity,
  cost control, 20, 59–61
  cost plan, 20
  cost report, 63
  duty in respect of final account, 90
  financial report, 61, 63–4
  interim valuation, 68
  monthly forecasts of final expendi-
    ture, 61
  rights, duties and liabilities, 9–10
  role in completion of contract after
    liquidation, 121, *et seq*

Temloc Ltd. v *Errill Properties – Court of
    Appeal 1987*, 105
tendering, selective, 12
tenders, 12
  analysis of, 20
time, extension of, 97, 98, 127

variations, 53, 60
  definition of, 51
  on sub-contract works, 55
  valuing of, 52–5